Empirical Research on Environmental Policies in China

Kiyoshi Fujikawa
Editor

Empirical Research on Environmental Policies in China

China Towards Decarbonization and Recycle Economy

Springer

Editor
Kiyoshi Fujikawa
Faculty of Economics
Aichi Gakuin University
Nagoya, Aichi, Japan

ISBN 978-981-99-5956-3 ISBN 978-981-99-5957-0 (eBook)
https://doi.org/10.1007/978-981-99-5957-0

This Springer imprint is published by the registered company Springer Nature Singapore Pte Ltd.
The registered company address is: 152 Beach Road, #21-01/04 Gateway East, Singapore 189721, Singapore

Preface

During the 40 years of "reform and door-opening policies" since the 1980s, China has achieved high economic growth by introducing certain market principles into its socialist economy. In 2010, China's GDP surpassed that of Japan to become the second largest in the world, and per capita GDP reached approximately $9000 in 2022, making China an expected member of the high-income nations in the near future.

Present-day China is advancing the sophistication of its economic and industrial structure with the aim of further development, but it is also facing many difficult problems. Japan has been said to have achieved in 100 years the economic growth that Europe and the United States took 200 years to achieve. This is something that should be appreciated, but it takes a certain amount of time for economic growth to achieve economic development including social growth. If that time of economic growth is compressed, distortion will occur. I, the editor of this book, realize this in the light of Japan's experience. China is trying to achieve in 50 years the economic growth that Japan took 100 years to achieve. Further compression of economic growth is thought to produce further distortions.

Distortion can be rephrased as challenges and one of such challenges is an environmental problem. The Chinese government was aware of environmental pollution caused by industrialization even before it adopted the reform and door-opening policies. Therefore, along with constructing an administrative organization for environmental protection, the enactment of environmental protection policies was promoted. However, the promotion of environmental policies requires companies and citizens to bear new costs. In the 1980s, shortly after the start of the reform and door-opening policies, the Chinese government did not have much leeway to spend domestic limited funds and resources on environmental conservation. In China at the time, as was the case in Japan during its high-growth period, environmental conservation hardly entered the economic logic.

One of the reasons for change in such attitude was the end of the Cold War. Until the 1980s, the concern of the international community was military security and economic growth that has supported military power. However, with the end of the

Cold War and the easing of military tensions, as global warming aggravated, the ozone layer ruptured, and biodiversity was threatened, the interest of the international community shifted to global environmental problems. The 1992 United Nations Conference on Environment and Development (Earth Summit) was held in such an atmosphere. The Chinese government must have felt the importance of environmental issues in international trends. China participated in the Earth Summit even though the Tiananmen Square crackdown in 1989 caused China to lose the confidence of the international community. In response to the Earth Summit, the Chinese government also formulated the Chinese version of Agenda 21, and "sustainable development" that balances economic development and environmental conservation has become one of China's central issues.

This book deals with climate change and waste management among the environmental problems facing China. Climate change is closely related to carbon dioxide emitted by fossil fuel consumption. China's carbon dioxide emissions overtook those of the United States in 2007 and China became the world's largest emitter. However, China promised to reduce carbon dioxide emissions (which were still expanding at that time) around 2030 and reduce emissions per GDP to about one-third of the current level in China's Nationally Determined Contribution (NDC) in the Paris Agreement of UNFCCC in 2015. Furthermore, President Xi Jinping surprised the international community in his speech at the 75th session of the United Nations General Assembly in September 2020, stating that China aims to achieve carbon neutrality by 2060. China has implemented low-carbon policies such as the establishment of a carbon dioxide emissions trading market and the promotion of renewable energy investment, and has achieved certain results. This book outlines what kind of global warming countermeasures are being implemented in China and analyzes the results of those policies from both micro and macro perspectives.

Moreover, the waste problem is becoming serious and the environmental pollution by the waste is also occurring in China. This book focuses on the automobile industry in view of the rapid increase in the number of end-of-life vehicles due to the progress of motorization in China. Developed countries have established a recycling system based on the principle of Extended Producer Responsibility (EPR) while China also in recent years has been promoting the establishment of laws related to recycling following the examples of developed countries. However, the design of the EPR system is complicated and there are many problems such as the recycling technology of companies and the environmental awareness of the people not reaching the level of developed countries. This book presents these issues while estimating the potential of recyclable resources obtained through automobile recycling. Another focal point of interest in this boom is the rapid increase in waste plastics in recent years. Because plastics are bulky, they place a heavy burden on waste disposal sites. For this reason, China embarked on an import ban on waste plastics at the end of 2017 that is still fresh in our memory. This book outlines such a waste plastic policy and estimates the environmental effect if recycling of plastic progresses in China.

China's environmental governance is a mixture of authoritarian regimes and market economy adjustments. In other words, while the policies set out by the central government are being implemented top-down by local governments and companies, price adjustments in markets are also functioning. While China is an economic superpower, it is also a mysterious country that is a "superpower" in terms of environmental problems. This book discusses environmental policies and their effects in such a hybrid country, China. As the editor, I would be very happy if this book could serve as a reference for readers who are interested in the environment and economy of such country.

The authors of this book are members of a mixed team of Japanese and Chinese researchers, but they are also research comrades who have spent time in the same laboratory. Needless to say, the participation of the Chinese researchers allowed us to have in-depth discussions on China's environmental policies. At the same time, I believe that the participation of Japanese researchers enabled us to discuss the evaluation of policies from a bird's-eye view that is a little more comprehensive. As the editor, I would like to thank all the authors for their cooperation.

Finally, I would like to express my gratitude to Mr. Yutaka Hirachi of Springer Japan for giving us the opportunity to publish this book, and to Ms. Kavitha Jayakumar of Springer Nature for her editorial efforts.

Nagoya, Aichi, Japan Kiyoshi Fujikawa

Contents

Editor and Contributors

About the Editor

Kiyoshi Fujikawa is a professor of economics at Aichi Gakuin University, Nagoya, Japan. He has over 30 years of research experience in the fields of environmental economics, development economics, international economics, and public finance/social security system in Nagoya University and Konan University. He got his PhD degree in economics at Kobe University. Besides universities, he also has served as a visiting researcher at the Economic and Social Research Institute (ESRI) of the Cabinet Office, the Institute of Population and Social Security Research (IPSS), and the Study Group on Economic Evaluation of Culture and Arts in the Agency for Cultural Affairs. He has served as President of the Pan-Pacific Association of Input-Output Studies (PAPAIOS) and as a board member of the Society of Environmental Economic Policy Studies (SEEPS) and the Japan Economic Policy Association (JEPA). Outside academia, he has worked for the Statistical Office in the Department of Economic and Social Affairs (DESA) in the United Nations (Chaps. 1, 2, 4, 5, 7, 9, and 10).

About the Contributors

Hikari Ban is a professor of economics at Kobe Gakuin University, Kobe, Japan. She got her PhD degree in economics at Kobe University, Japan. Her major is international economics with computable general equilibrium analysis, and her research interests include trade and environmental load, trade and prices of primary inputs, and international emissions trading system (Chap. 9).

Tadashi Hayashi is an associate professor of economics at the University of Shiga Prefecture, Japan. He got his PhD degree in economics at Kyoto University, Japan. His major is environmental economics in the fields of recycling policy and industries, trade and environmental policy spillover, and design of GHG emission trading system (Chap. 3).

Yiyi Ju is a lecturer of economics at Waseda Institute for Advanced Studies (WIAS), Waseda University, Tokyo, Japan, after serving as a researcher at the University of Tokyo. She got her PhD degree in international development at Nagoya University. Her major is environmental economics, especially in the field of environment-related industrial technological innovation and environmentally friendly consumer behavior (Chaps. 1, 7, and 8).

Yang Li is Associate Professor of economics at Zhongnan University of Economics and Law, Wuhan, China. She got her PhD degree in international development at Nagoya University, Japan, after working for Mazda Motor Corporation in Hiroshima. Her research fields include green development of automobile industry and transboundary embodied energy flow (Chaps. 2 and 8).

Xi Lu is a lecturer of economics at Hunan University of Finance and Economics, Changsha, China. She got her PhD degree in international development at Nagoya University, Japan, after getting her master's degree from the University of Shiga Prefecture, Japan. Her research interest includes recycling policy and industries, and waste disposal and management (Chap. 10).

Jiayang Wang is a lecturer of economics at the Faculty of Economics, Aichi Gakuin University, Nagoya, Japan, after serving as a researcher at the University of Tokyo and Renewable Energy Institute. He got his PhD degree in international development at Nagoya University after getting his master's degree from the University of Shiga Prefecture, Japan. His research fields include environmental policy and renewable energy introduction (Chaps. 1 and 6).

Takatoshi Watanabe is a professor of economics at Aichi Gakuin University, Nagoya, Japan. He got his master’s degree of engineering at Toyohashi University of Technology and another master’s degree of economics at Tezukayama University. His major is application of input-output analysis to various economic issues in the fields of international economics, environmental economics, public finance, and disaster economics. He has been a board member of the Japan Economic Policy Association (JEPA) (Chap. 5).

Zuoyi Ye is an associate professor of economics at Shanghai University of International Business and Economics, Shanghai, China, after working for Applied

Research Institute in Tokyo. He got his PhD degree in international development at Nagoya University, Japan. His major is development economics, especially in the fields of industrial structural change, international trade, and China's foreign aids. He has been a board member of the China Input-Output Association (CIOA) for many years (Chap. 4).

Part I
Environmental Policies in China

Chapter 1
Climate Policies in China: Renewable Energy Introduction and National Emissions Trading Scheme

Jiayang Wang, Yiyi Ju, and Kiyoshi Fujikawa

Contents

Keywords Renewable energy · Feed-in tariff · Emission trading scheme · Carbon price

Abbreviations

CCER	Chinese Certified Emission Reduction
CEA	Carbon Emission Allowance
CO_2	Carbon Dioxide

J. Wang (✉) · K. Fujikawa
Aichi Gakuin University, Nagoya, Aichi, Japan
e-mail: jywang@dpc.agu.ac.jp; fujikawa@dpc.agu.ac.jp

Y. Ju
Waseda University, Shinjuku City, Tokyo, Japan
e-mail: juyiyi@aoni.waseda.jp

K. Fujikawa (ed.), *Empirical Research on Environmental Policies in China*,
https://doi.org/10.1007/978-981-99-5957-0_1

COP	Conference of the Parties to the United Nations Framework Convention on Climate Change
ETS	Emission Trading Scheme
EU-ETS	European Union Emissions Trading Scheme
FIT	Feed-in Tariff
GHG	Greenhouse Gas
LCOE	Levelized Cost of Electricity
NDRC	National Development and Reform Commission
RE	Renewable Energy

1 Introduction

At the 21st Conference of the Parties to the United Nations Framework Convention on Climate Change (COP21) held in Paris in 2015, participating countries agreed to implement global warming countermeasures on a global scale. Since then, as the world has accelerated the movement toward decarbonization, China has set a goal of achieving carbon neutrality by 2060. To achieve this goal, the Chinese government has implemented various policies. In this chapter, we explain China's renewable energy (RE) introduction policy and carbon dioxide (CO_2) emissions trading system as decarbonization policies.

Coal is the main source of energy consumption in China, accounting for approximately 70% of primary energy consumption. Coal emits more CO_2 per calorific value than any other fossil fuel. This is partly why China is the world's largest CO_2 emitter, although its economy is smaller than that of the United States. Since most of China's coal consumption is used for power generation, shifting from thermal power generation to RE generation and introducing a CO_2 emissions trading system in the power sector are effective measures to reduce coal consumption.

The Chinese government enacted the Renewable Energy Law, which requires electric power companies to purchase RE power, and promoted a series of policies to encourage the introduction of RE power generation, including a "Concession bidding system for renewable power" and a "feed-in tariff" (FIT) system for RE power. As a result, China has become the world's largest user of RE. Section 2 introduces the policies that promote the expansion of RE use in China.

China also started a unified national emissions trading scheme (ETS) for the electric power industry on July 16, 2021. This scheme was launched 8 years after China's pilot carbon market was established in Shenzhen in June 2013 and 6 years after the State Council announced its intention in September 2015 to establish a unified national carbon emissions market. In Sect. 3, the pilot projects for the ETS in nine Chinese cities and regions (Shenzhen, Shanghai, Beijing, Guangdong, Tianjin, Hubei, Chongqing, Sichuan, and Fujian) are outlined. We also describe the characteristics of the national unified ETS for electric power and examine the market performance in 2021 in Sect. 3. Section 4 discusses potential future issues regarding RE policies and the ETS in China.

2 Policies for RE Energy Promotion in China

2.1 Status of the Introduction of RE

Figure 1.1 depicts the introduction status of RE in China. The line graph indicates the share of RE in the total power generation. It was only 16% in 2011, but it doubled to 29% in 2020. The bar graph indicates the composition of RE. RE comprises hydropower, wind power, solar power, and biomass, among which hydropower has the largest share, accounting for more than half of RE in 2020. However, the expansion rate of wind and solar power generation after 2011 is extremely large. If you take a look at their share in renewable energies, the share of wind power has tripled from 6.5% to 21.1%, and that of solar power increased from 0.0% to 11.8%.

2.2 Introduction of a Concession Bidding System for RE

China's RE industry development policy began in the 1990s for wind power generation and the 2000s for solar power generation. The RE Law, which came into effect in 2006, defined RE and clarified its role in the Chinese economy. The law also obliged power transmission and distribution companies to purchase the entire amount of electricity generated by RE generators. With the enforcement of this law, China's RE power generation has increased dramatically.

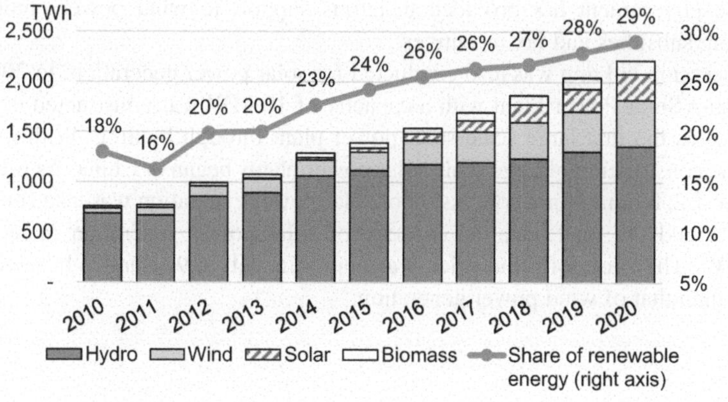

Fig. 1.1 Power generation and rate of China's RE. Source: Created by the authors using data from National Bureau of Statistics (2021)

Table 1.1 Conditions and results of concession bidding for wind power

Year	Equipment domestic production rate standards	Region	Capacity (MW)	Contract price (yuan/kWh)
2003	Over 50%	Guangdong	200	0.501
		Jiangsu		0.436
2004	Over 50%	Inner Mongolia	400	0.382
		Jiangsu		0.519
		Jilin		0.509
2005	Over 70%	Jiangsu	450	0.488
		Gansu		0.462
		Shandong		0.600
2006	Over 70%	Inner Mongolia A	700	0.420
		Inner Mongolia B		0.466
		Hebei		0.501
2007	Over 70%	Inner Mongolia A	950	0.468
		Inner Mongolia B		0.522
		Hebei		0.551
		Gansu		0.521

Source: Created by the authors based on the studies by Jiang and Shi (2006) and Liu (2010)

In addition, the National Development and Reform Commission (NDRC) have started the "concession bidding system for RE." This system calls for bidders for development rights for power generation projects of 100 MW or more in areas where natural energy resources are rich. Further, the system focuses on reducing power generation costs and increasing the domestic production rate of power generation facilities.

As presented in Table 1.1, the concession bidding system for wind power began in 2003. The capacity of wind power through concession bidding has reached 2700 MW. The contract price was approximately 0.5 yuan/kWh. Since 2006, the Chinese government has provided generous support to wind power companies, including subsidies and tax incentives.

Concession bidding was also conducted for solar power generation. In 2009, the Dunhuang Solar Power Plant with a capacity of 10 MW was constructed in Gansu Province as the first large scale solar power plant through bidding. With this as a start, the construction of large scale solar power plants began in China. As presented in Table 1.2, bidding for a concession of solar power generation was also conducted in 2010, and the total installed capacity of solar power generation increased to 290 MW. The successful bid price was approximately 0.9 yuan/kWh, which was higher than that of wind power generation.

Table 1.2 Conditions and results of concession bidding for solar power

Year	Region	Capacity (MW)	Contract price (yuan/kWh)
2009	Gansu	10	1.0900
2010	Shaanxi	20	0.8687
	Qinghai A	30	0.7288
	Qinghai B	20	0.8286
	Gansu A	20	0.8265
	Gansu B	20	0.7803
	Gansu C	20	0.8099
	Inner Mongolia A	20	0.8847
	Inner Mongolia B	20	0.7978
	Inner Mongolia C	20	0.8444
	Ningxia	30	0.9791
	Xinjiang A	20	0.7388
	Xinjiang B	20	0.9317
	Xinjiang C	20	0.9907

Source: Created by the authors based on the study of Hu (2011)

2.3 Introduction of the FIT System

FIT is a system that obligates power transmission companies to purchase power generated by RE for a certain period at a price determined by the government. FITs for wind and solar power started in 2009 and 2011, respectively, ending the bidding system for RE development rights. Regarding the fixed purchase price, the purchase price for wind and solar power generation is set by region, and the purchase period for both is 20 years.

The Renewable Energy Law Amendment Bill enacted in 2010 stipulates that power transmission and distribution companies should share the increased cost of purchasing RE power generation among electricity consumers nationwide. With this, the collection system of the RE power generation promotion levy started.

The purchase price changes every year, Table 1.3 presents the purchase price for wind-generated power as of 2019, and Table 1.4 presents the purchase price for photovoltaic power as of 2019. The purchase price is categorized according to the abundance of RE resources and the construction cost in each region. Four resource zones were established for wind power generation, and the purchase price was set low in areas with abundant wind power resources. Further, three categories of resource zones were established for solar power generation, and the purchase price was set low in areas rich in photovoltaic resources.[1] The purchase price of solar power is slightly higher than that of wind power.

[1] There are some "local production for local consumption" type of projects in solar power generation as an anti-poverty measure in poor rural areas. In such cases, the resource zone was the same as the general solar power generation, but the purchase price was set higher.

Table 1.3 FITs by resource zone for wind power (2019)

Resource zone	Purchase price	Target area
1st	0.34 Yuan/kWh	Inner Mongolia: Regions other than those in the second zone Xinjiang: Urumqi, Ili, Kelamayi, and Shihezi
2nd	0.39 Yuan/kWh	Inner Mongolia: Chifeng, Tongliao, Xin'an, and Hulunbuir Hebei: Zhangjiakou and Chengde Gansu: Jiayuguan and Jiuquan
3rd	0.43 Yuan/kWh	Jilin: Baicheng and Songyuan Heilongjiang: Jixi, Shuangyashan, Qitaihe, Suihua, Yichun, and Daxin'anling Gansu: Regions other than those in the second zone Xinjiang: Regions other than those in the first zone Ningxia
4th	0.52 Yuan/kWh	Regions other than those in the first, second, and third zones

Source: Created by the author based on the NDRC (2019a)

Table 1.4 FITs by resource zone—solar power (2019)

Resource zone	purchase price	Target area
1st	0.40 Yuan/kWh	Ningxia, Qinhai: Haixi Gansu: Jiayuguan, Wuwei, Zhangye, Jiuquan, Dunhuang, and Jinchang Xinjiang: Hami, Tacheng, Altay, and Kelamayi Inner Mongolia: Regions other than the second zone
2nd	0.45 Yuan/kWh	Beijing, Tianjing, Heilongjiang, Jilin, Liaoning, Sichuan, and Yunnan Inner Mongolia: Chifeng, Tongliao, Xin'an, and Hulunbuir; Hebei: Chengde, Zhangjiakou, Tangshan, and Qinhuangdao Shanxi: Datong, Shuozhou, and Yizhou Shaanxi: Yulin and Yan'an Qinghai: Regions other than those in the first zone Gansu: Regions other than those in the first zone Xinjiang: Regions other than those in the first zone
3rd	0.55 Yuan/kWh	Regions other than those in the first and second zones

Source: Created by the authors based on data from the NDRC (2019b)

2.4 *End of FIT System*

The levelized cost of electricity (LCOE)[2] of wind power in China in 2019 was about US$50/MWh, which is about half of that in 2014, and the LCOE of solar power in 2019 was about US$ 43/MWh, which is about a quarter of that in 2014. However, the LCOE of coal-fired power generation in China in 2019 was US $50–72/MWh, which is at the same level as that of wind and solar power (Wang (2020)). As a result, the FIT system for power generated by RE has lost its meaning, and the FIT system for large scale wind and solar power generation finished by 2021.[3]

After that, the Chinese government started a new system called the "grid parity project." Grid parity projects are projects in which wind and solar power plants sell electricity at benchmark prices for coal-fired power generation in the same region for more than 20 years without receiving subsidies. Participants in this project can enjoy preferential treatment, such as priority rights in power transmission, purchase of all generated electricity, issuance of a "Green Power Certificate," guarantee of grid connection, and financial support. The first solicitation for grid parity projects (4.5 GW) was made in May 2019. The breakdown was 4.5 GW for onshore wind power, 14.8 GW for solar power, and 1.5 GW for distributed power generation ("local production for local consumption" type).

After the end of the FIT, the interest in RE power policy shifted from quantitative expansion to market efficiency. In 2021, a new product called "green power" was introduced in the power trading market, adding environmental value to RE power. Suppliers of RE electricity are required to obtain the above-mentioned "Green Power Certificates," which are purchased for firms to reduce CO_2 emissions. In addition, as already mentioned, as RE power fluctuates depending on the weather, the supply of RE power requires adjustment power to stabilize the electricity in the grid. The Chinese government established the "Ancillary Service Market" (CNCTST (2020)) to trade electric power adjustment means, such as power storage facilities or pumped storage power generation. This has made the RE market more efficient.

3 ETS in China

3.1 *Industries and Companies Covered by the ETS*

As a market-based climate policy instrument, ETS provides economic incentives for its covered entities to achieve emission mitigation. Currently, 26 ETSs are in force,

---

[2]LCOE means the average cost per unit of power generation. It is calculated based on the total costs necessary for power generation, such as construction costs, operation and maintenance costs, and fuel costs as well as profits and the estimated power generation during the operation period.

[3]Subsidization policies now continues for offshore wind power and distributed power generation, but they are scheduled to phase out by 2030.

including EU-ETS, Korea ETS, and Japan ETS (Tokyo, Saitama) (ICAP (2022)). The catalyst for creating an ETS in China was the 15th Conference of the Parties at the United Nations Framework Convention on Climate Change (COP15) held in Copenhagen in December 2009. At COP15, the Chinese government announced a national goal of reducing the country's carbon emissions per unit of GDP by 40%–45% of the 2005 levels by 2020. The carbon markets were set up to achieve this goal in 2020.

In China, the government set up pilot programs for a certain systemic reform in certain areas before the implementation of nationwide systemic reforms. One example of such pre-implementation processes is the recent reform of indirect taxes in the 2010s. Until 2011, China's indirect tax system entailed a business tax and a value-added tax. Goods were subject to the value-added tax, whereas the business tax was levied on services. The dual existence of these taxes had many adverse effects, so the two taxes were integrated to form a single value-added tax. The pilot project to integrate them into one value-added tax was introduced in Shanghai on January 01, 2012, followed by Beijing, Jiangsu, and other provinces, and then expanded nationwide in 2013. However, until 2013, the services subject to the tax did not include rail transport, telecommunications, or real estate. The service industries subject to the tax were gradually increased, and all service industries nationwide came under the value-added tax system in 2016.

The ETS also started in the form of pilot projects in several rural areas. In October 2011, the NDRC issued the "Notice on the Implementation of ETS Pilot Projects." In 2013, the ETS pilot project was introduced in Shenzhen (June 18, 2013), Shanghai (November 26, 2013), Beijing (November 28, 2013), Guangdong (December 19, 2013), and Tianjin (December 26, 2013). In 2014, other ETS pilot projects were launched in Hubei (April 2, 2014) and Chongqing (June 19, 2014). After that, another pilot market was opened in Sichuan (December 16, 2016) and Fujian (December 18, 2016), making the total number of pilot markets nine.

As presented in Table 1.5, in each of these ETS pilot projects, the criteria for the covered industries and firms within these industries are quite different. Shenzhen and Shanghai, which have the widest range of target industries, have the highest number of industries, whereas Chongqing regulates only heavy industries. Every area regulates its electricity and heat supply and steel industries. Many areas also regulate the chemical and petrochemical industries. Shanghai is the only one regulating the construction sector. In terms of the threshold criteria for companies located in the target industries, Shenzhen has the lowest standard emission (i.e., 3000 t-CO_2). However, with baseline emissions of 60,000 coal-equivalent tons (tce) of energy use (approximately 156,000 t CO_2-equivalent) until 2016, Hubei Province's market had the highest baseline CO_2 emissions. This figure is the highest of all the markets even after the baseline level was reduced to 10,000 tce (about 38,000 t CO_2-equivalent) in 2017. The next largest markets in terms of baseline CO_2 emissions

Table 1.5 Covered industries and firms for each ETS pilot project

Area	Target industries	Remarks (as of 2018)
Beijing	Electricity and heat supply, cement, petrochemicals, manufacturing, transportation (buses, subways, etc.), services, and other industrial sectors	· Those with average annual emissions of 10,000 t-CO_2 or more from 2009 to 2011 · From 2016, those with emissions of 5000 t-CO_2 or more in 2014 were added
Tianjin	Electricity and heat supply, steel, chemicals, petrochemicals, and natural gas extraction	· Those with average annual emissions of 20,000 t-CO_2 or more from 2009 to 2015 · From 2016, those with emissions of 10,000 t -CO_2 or more were also included
Shanghai	Electricity and heat supply, steel, petrochemicals, chemicals, nonferrous metals, construction materials, spinning and weaving, paper manufacturing, rubber, synthetic textiles, transportation (air transportation, airports, shipping, ports, and subways), commerce, hotels, and finance	· Industrial sectors: Those with emissions of 20,000 t-CO_2 or more in either 2010 or 2011 · Nonindustrial sectors: Those with emissions of 10,000 t-CO_2 or more in either 2010 or 2011
Shenzhen	Twenty-six industrial sectors, including electricity and heat production, processing, manufacturing, transportation (ports, buses, and subways), and large public buildings	· Industries that have had emission levels of 3000 t-CO_2 or more during any of the years from 2009 to 2011 · Operators of large public facilities of 10,000 square meters or more
Chongqing	Heavy industries	· Industries that have had emission levels of 20,000 t-CO_2 or more during any of the years from 2008 to 2012
Guangdong	Electricity and heat supply, steel, cement, petrochemicals, paper manufacturing, and aviation (air transportation and airports)	· Those with emissions of 20,000 t-CO_2 or more in either 2011 or 2012
Hubei Province	Electricity and heat supply, steel, automobiles, nonferrous metals, glass, cement, chemicals, petrochemicals, food, synthetic textiles, paper manufacturing, pharmaceuticals, etc.	· Those with emissions of 60,000 t-CO_2 or more in 2010 or 2011 · From 2017, industries that have had energy consumption levels of 10,000 tce or more in any year from 2014 to 2016 were included
Fujian Province	Electricity, steel, chemical industry, petrochemicals, nonferrous metals, aviation (air transportation and airports), construction materials, paper, and ceramics	· Industries that have had energy consumption levels of 10,000 tce or more in any year from 2013 to 2015 were included
Sichuan Province	Electricity, steel, chemicals, nonferrous metals, paper, aviation (air transportation and airports), and construction materials	· Industries with emission levels of 26,000 t-CO_2 or more in 2016 or 2017

Source: Created by the authors based on ICAP (2022)

are Sichuan and Fujian, at 26,000 tons.[4] The number of companies subject to the criteria started to increase in 2017, as the criteria were lowered from 10,000 t-CO_2 to 5000 t-CO_2 in the Beijing pilot project and from 20,000 t-CO_2 to 10,000 t-CO_2 in the Tianjin pilot project.

3.2 Setting Emissions Allowances for ETS Pilot Projects

The allocation of emission allowances becomes complicated when the scheme covers multiple areas and industries (Demailly and Querion (2006)). Emission allowances for the ETS pilot projects vary from area to area and industry to industry because the following factors were considered: (1) regional CO_2 reduction targets and targets for related sectors, (2) air pollution control targets, and (3) regional economic and industrial development targets. The differences are also because economic and air pollution conditions vary from region to region (Jiang et al., 2016), the CO_2 reduction targets (vs. 2015) in the 13th Five-Year Plan (2016–2020) vary from region to region, and the targets are not for the total CO_2 emissions but for CO_2 emissions per unit of GDP, making the context of China more difficult to understand. Although the aggregate CO_2 reduction target per unit of GDP for China is 18.0%, it is higher (20.5%) in the more developed areas of Beijing, Tianjin, Hebei, Shanghai, Jiangsu, Zhejiang, Shandong, and Guangdong. Furthermore, this target is 19.5% in Fujian, Jiangxi, Henan, Hubei, Chongqing, and Sichuan. In Shanxi, Liaoning, Jilin, Anhui, Hunan, Guizhou, Yunnan, and Shaanxi, the reduction target is 18.0%, which is the same as the national target. However, in the less developed northwestern provinces of Inner Mongolia, Heilongjiang, Guangxi, Gansu, and Ningxia, the reduction target, which is 17%, is below the national target. In Hainan, Tibet, Qinghai, and Xinjiang, the reduction target is 12.0%, which is far below the national target.

The emission allowances are calculated using either the benchmarking method or the grandfathering method. The benchmarking method involves multiplying the actual corporate activity by a coefficient to obtain the emission rate or any other metric on the industry's top players. The grandfathering method involves applying a certain reduction rate to the base-year emissions volume (average annual emissions during the past several years).

[4]One ton of coal-equivalent energy use corresponds to approximately 2.6 t-CO_2 in terms of CO_2 emissions.

3.3 Characteristics of the Unified National ETS

By incorporating the experience gained from the regional pilot projects, the unified national ETS was launched in July 2021 and started in the electricity supply sector and may be extended to cement, petrochemicals, nonferrous metals, steel, etc. Consequently, China now has a national market and nine local markets operating simultaneously. The national ETS has two locations—the Unified National Carbon Emission Rights Registration and Exchange in Wuhan and the Unified National Carbon Emission Rights Exchange in Shanghai. A carbon registration system and a trading system were established and are being operated by the China Hubei Provincial Carbon Emissions Exchange and the Shanghai Environment and Energy Exchange, respectively.

Currently, the National ETS only covers the electric power sector. Moreover, it covers a total of 2163 power plants that have extremely high annual CO_2 emission levels, i.e., more than 26,000 tons. The benchmark values for CO_2 emissions quotas vary depending on the type of power plant, which are 0.877 t-CO_2/MWh for coal-fired power plants of 300 MW class and above, 0.979 t-CO_2/MWh for coal-fired power plants of 300 MW class and below, and 0.392 t-CO_2/MWh for gas turbines. These figures are then further adjusted based on the cooling method or the volume of heat supplied. The initial allocations were carried out without charge. The preliminary allocation (initial allocation) of emissions allowances is based on 70% of the actual electricity supply in the 2018 fiscal year. The allocation of emissions allowances was then adjusted in accordance with the actual electricity supply in 2019 and 2020.

There are two methods for trading emissions allowances—negotiated trades and bidding. Furthermore, there are two types of negotiated trades—listed negotiated trades and large-lot negotiated trades. For listed negotiated trades, the size of the trade is 100,000 t-CO_2 or less, and the trading price is set within a range of 10% above or below the closing price of the previous trading day. Large-lot negotiated trades are trades wherein the size of the trade is 100,000 t-CO_2 or more, and the trading price is set within a range of 30% above or below the closing price of the previous trading day.

Currently, Carbon Emission Allowances (CEAs) are traded by China's national ETS (Cong & Wei, 2010). The allocation approaches of the allowances have been debated for a long time—since market-based carbon pricing started (Flues & van Dender). These allowances have been allocated to each company. Chinese Certified Emission Reductions (CCERs) will also be traded in the near future. CCERs are carbon offset credits issued for emission reduction in voluntary carbon emission reduction projects. They are equivalent to J-credits (formerly known as domestic credits) in Japan.

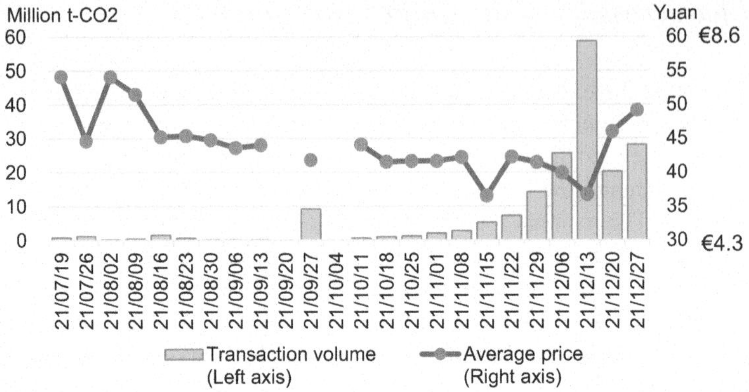

Fig. 1.2 Trading volume and prices in China's ETS. Source: Created by the authors based on data from the National Carbon Market Transaction Data Overview. Available at: <http://www.tanjiaoyi.com/tanhangqing/>

3.4 ETS Price Movements

According to the National Carbon Market Transaction Data Overview, the total volume of CEA transactions from the opening of the national ETS in July 2021 to the end of the year was 179 million tons, with a total transaction value of RMB 7.661 billion. From this figure, the average price of CO_2 is RMB 42.8 per ton, which is approximately 770 yen at a rate of 18 yen per RMB.

The restricted period is 1 year, and covered companies must achieve their emissions quota by year-end. Companies that exceed their quota must purchase emissions allowances from the ETS or purchase offset credits (a "cap and trade" system). Figure 1.2 depicts the weekly volumes and prices traded in the national ETS. Regarding price movements, the price exceeded RMB 50 per ton in the first few weeks of trading but gradually declined thereafter, falling to RMB 40 per ton in December 2021. However, at the end of the year, when the covenant period had almost ended, the price rose to 50 RMB per ton. This is still considerably lower than that in the European trading market (European Union-ETS (EU-ETS)). The price in Europe has increased sharply since December 2020—when the EU raised its greenhouse gas (GHG) reduction target. Therefore, as of September 1, 2021, it was more than 60 euros per ton of CO_2 (7800 yen at a rate of 130 yen per euro).

Regarding the trading volume, except for the week before China's National Day, the weekly trading volume did not reach ten million tons until late November. However, as the end of the covenant period approached, the volume quickly increased, and during the week of December 13, the trading volume exceeded 50 million tons.

4 Future Issues

4.1 RE

In 2021, the Chinese government announced the "Circular of the State Council on an action plan for peaking carbon emissions before 2030." Table 1.6 summarizes the five-year action target related to RE. As of 2020, the share of RE in primary energy consumption was 15.9%, but the plan aims to increase this ratio to 20% by 2025 and 25% by 2030. This means that more than half of the new increase in primary energy demand will be covered by RE.

Regarding the introduction of RE, the target is set to build 80 GW of new hydroelectric power plants and expand the total installed capacity of wind and solar power generation to over 1200 GW by 2030. It is also decided to develop large scale offshore wind power generation. Although the national target for offshore wind power generation has not been announced, the total individual targets of several provinces that are positive about the development of offshore wind power generation is to reach approximately 60 GW by 2025.

If the above introduction targets are achieved, the total capacity of RE power generation facilities will greatly exceed the 2021 figure of 1259 GW of thermal power generation in 2030, and RE will become China's main power source.[5] However, in addition to the introduction of a large amount of electricity from RE sources, it is necessary to improve the flexibility and stability of the power grid. Therefore, the "long-term target for RE" set a goal of installing 30 GW of storage capacity by 2025 and increasing the total installed capacity of pumped storage to 120 GW by 2030.

Table 1.6 China's long-term target for RE

	2021–2025	2026–2030
RE rate in primary energy consumption	20%	25%
RE rate in new primary energy consumption	50%	–
New hydropower installed capacity	40 GW	40 GW
Total installed capacity of wind and solar power generation	–	1200 GW
New power storage installed capacity	30 GW	–
Total installed capacity of pumped-storage power generation	–	120 GW
RE power generation	3.3 Tri. kWh	–
Ratio of RE power generation	33%	–

Source: Created by the authors based on papers by State Council (2021) and National Development and Reform Commission (NDRC) (2022a)

[5]Peking University Institute of Energy (2021) predicts that CO_2 emissions from the power sector will peak in 2025.

The emphasis of policies that are aimed at increasing the proportion of RE is shifting from expanding the introduction of RE to efficient use of RE. As already mentioned, subsidies for large scale solar power and onshore wind power generation have been abolished, and currently, only offshore wind power generation is eligible for subsidies. It was decided to promote the decarbonization of transportation and the use of hydrogen as an energy carrier to improve the efficiency of using RE, the development of RE storage facilities, and the strengthening of power grids.

In the field of transportation, the government is strongly promoting the introduction and expansion of electric vehicles (EV) and fuel cell vehicles (FCV) because they use renewable power. The central government has implemented measures such as providing subsidies to consumers and reducing or exempting automobile acquisition taxes to support the purchase of new energy vehicles (Rui et al. (2021)). Although the central government's preferential treatment will phase out in the future, local governments are expected to continue to provide their own subsidies. In addition, the government has developed transportation infrastructure, such as charging stations for EVs and hydrogen filling stations for FCVs, to promote the use of EVs and FCVs (Rui et al. (2021)). Moreover, the National Development and Reform Commission (NDRC) (2022b) obliged new detached houses built after 2022 to be designed so that EV charging facilities can be added after construction and called for existing residential complexes and commercial facilities to install EV charging facilities.

Incidentally, the use of green hydrogen[6] is attracting attention as a method of effective use of RE. Hydrogen was considered as energy for the first time in the "Energy Law (Provisional)" enacted in 2020 (National Energy Administration (2020)). However, expanding the use of hydrogen does not necessarily lead to decarbonization as most hydrogen is currently made from fossil fuels (so-called gray hydrogen).[7] Therefore, green hydrogen is the basis for future hydrogen production in the "Hydrogen industry Development Medium- to Long-Term Plan 2021–2035" announced in 2022 (National Development and Reform Commission (NDRC) (2022c)). However, the produced hydrogen does not have a green or gray color, so users cannot distinguish between them. Therefore, the government launched China's first "green hydrogen certification system" in 2020 to support the green hydrogen industry (China Hydrogen Alliance (2020)). The government's support policy for green hydrogen has been substantialized.

[6]Green hydrogen is hydrogen produced by electrolyzing water using electricity from RE sources.

[7]Currently, there are two main hydrogen production methods—reforming hydrogen from fossil fuels (e.g., steam reforming) and recovering hydrogen generated from coke ovens, etc. Hydrogen processed by such methods is called gray hydrogen as these production processes emit CO_2.

4.2 ETS

The Chinese government has a high level of confidence in the ETS. The ETS will provide economic incentives to reduce GHG emissions, reduce the cost of reducing GHG emissions, and promote green technology innovation and industrial investment. At present, China's ETS only covers the electric power industry. However, there are plans to create ETSs for other energy-intensive industries, such as cement, petrochemicals, nonferrous metals, and steel. Whether these markets will emerge as industry-specific carbon markets or integrated carbon markets that involve multiple industries is still being considered.

There is another issue that requires addressing and reconciliation in the future. This issue pertains to the coexistence of CO_2 reduction targets for businesses and provinces. Province-specific CO_2 reduction targets must also be strictly met.[8] In many cases, the production area (i.e., the area of energy consumption) and the final consumption area are different; the electric power industry is a typical example. Some people believe that individuals in areas where goods and services are consumed should also take responsibility for CO_2 emissions. The issue of coordination between the production and consumption areas for goods and services is also likely to arise.

It is interesting that while China, a socialist country, is emphasizing price mechanism, it is also ironic that Japan, a market economy, does not have a national-level carbon market. However, in December 2021, Japan's Ministry of Economy, Trade, and industry also launched the "Working Group on the Improvement of the Environment for Appropriate Use of Carbon Credits to Achieve Carbon Neutrality." Although Japan has carbon credit schemes such as J-credits, the trading of credits in these schemes are arm's length transactions that do not provide price signals. As carbon pricing (a policy to change the behavior of emitters by putting a price on carbon) is becoming increasingly common around the world, Japan can also create a system to facilitate trading in carbon credits.

Acknowledgements This work was supported by JSPS KAKENHI Grant Numbers JP19K12459372 and JP21H04941.

References

China National Committee for Terminology in Science (CNCTST) (2020) Chinese Terms in Electric Power. China Science Publishing & Media. (in Chinese)

China Hydrogen Alliance (2020) Standard and evaluation of low-carbon hydrogen, clean hydrogen and renewable hydrogen. (in Chinese). http://www.ttbz.org.cn/StandardManage/Detail/42014/

Cong R, Wei Y (2010) Potential impact of (CET) carbon emissions trading on China's power sector: A perspective from different allowance allocation options. Energy 35(9):3921–3931. https://doi.org/10.1016/j.energy.2010.06.013

[8] See Reuters' article dated September 16, 2021.

Demailly D, Querion P (2006) CO2 Abatement, Competitiveness and Leakage in the European Cement Industry Under the EU ETS: Grandfathering Versus Output-Based Allocation. In: Grubb M, Neuhoff K (eds) *Emissions Trading and Competitiveness: Allocations, Incentives and Industrial Competitiveness under the EU Emissions Trading Scheme*, chapter 5. Routledge, pp 93–114

Hu R (2011) Development and thoughts of PV power concession bidding projects. Sol Energy 2011(5):10–15. (in Chinese). https://doi.org/10.3969/j.issn.1003-0417.2011.05.003

International Carbon Action Partnership (ICAP) (2022) Emissions Trading Worldwide: Status Report 2022. International Carbon Action Partnership. https://icapcarbonaction.com/en/publications/emissions-trading-worldwide-2022-icap-status-report

Jiang L, Shi P (2006) Implementation Situation of Wind Power Concession Bidding Projects in China and Analysis. Electric Power Technol Econ 18(4):1–3. (in Chinese). https://doi.org/10.3969/j.issn.1674-8441.2006.04.002

Jiang J, Xie D, Ye B, Shen B, Chen Z (2016) Research on China's cap-and-trade carbon emission trading scheme: Overview and outlook. Appl Energy 178:902–917. https://doi.org/10.1016/j.apenergy.2016.06.100

Liu L (2010) Comprehensive Analysis of China's Fifth of Wind Power Concession Bidding. China Commerce 204:322–322. (in Chinese). https://d.wanfangdata.com.cn/periodical/ChlQZXJpb2RpY2FsQ0hJTmV3UzIwMjIxMTE1Eg5RSzIwMTAwMjA3NTExORoIN3I4aXZyeDU%3D

National Bureau of Statistics (NBS) (2021) Chinese Energy Statistical Yearbook 2020. China Statistics Press. (in Chinese)

National Development and Reform Commission (NDRC) (2019a) Notice on improving the wind power feed-in tariff policy. https://chinaenergyportal.org/en/notice-on-improving-the-wind-power-feed-in-tariff-policy/

National Development and Reform Commission (NDRC) (2019b) Circular on improving the feed-in tariff mechanism for PV power generation. https://chinaenergyportal.org/en/circular-on-improving-the-feed-in-tariff-mechanism-for-pv-power-generation/

National Development and Reform Commission (NDRC) (2022a) 14th five-year plan for renewable energy development. https://chinaenergyportal.org/en/14th-five-year-plan-for-renewable-energy-development/

National Development and Reform Commission (NDRC) (2022b) Opinions on further improving the guaranteed servicing capability of electric vehicle charging infrastructure. https://chinaenergyportal.org/en/opinions-on-further-improving-the-guaranteed-servicing-capability-of-electric-vehicle-charging-infrastructure/

National Development and Reform Commission (NDRC) (2022c) Development plan of hydrogen energy industry for the 2021–2035 period. (in Chinese). https://www.gov.cn/xinwen/2022-03/24/5680975/files/6b388f7c324a4b1db0b30dc6f52b7e02.pdf

National Energy Administration (2020) Energy Law of the People's Republic of China (draft). https://chinaenergyportal.org/en/energy-law-of-the-peoples-republic-of-china-draft-for-comments/

Peking University Institute of Energy (2021) Pathways and Policy for Peaking CO_2 emissions in China's Power Sector. https://www.ccetp.cn/newsinfo/2846086.html

Rui K, Zhuang F, Kai F, Kang L, Jia W (2021) Moving from subsidy stimulation to endogenous development: A system dynamics analysis of China's NEVs in the post-subsidy era. Technol Forecast Soc Chang 168:1–10. https://doi.org/10.1016/j.techfore.2021.120757

State Council (2021) Circular of the State Council on An Action Plan for Peaking Carbon Emissions Before 2030. https://chinaenergyportal.org/en/circular-of-the-state-council-on-an-action-plan-for-peaking-carbon-emissions-before-2030/

Wang J (2020) Energy Structural Transformation and Renewable Energy Growth in China, Renewable Energy Institute. (in Japanese). https://www.renewable-ei.org/pdfdownload/activities/ChinaReport_JP.pdf

Chapter 2
Introduction of Extended Producer Responsibility in China

Yang Li and Kiyoshi Fujikawa

Contents

Keywords End-of-life vehicles · Extended producer responsibility · Electric vehicle

Abbreviations

ASR	automobile shredder residue
BEV	battery electric vehicles
ELV	End-of-Life Vehicles
EPR	Extended Producer Responsibility
METI	Ministry of Economy, Trade and Industry

Y. Li (✉)
Zhongnan University of Economics and Law, Wuhan, Hubei, P.R.China
e-mail: liyang@zuel.edu.cn

K. Fujikawa
Aichi Gakuin University, Nagoya, Japan
e-mail: fujikawa@dpc.agu.ac.jp

K. Fujikawa (ed.), *Empirical Research on Environmental Policies in China*,
https://doi.org/10.1007/978-981-99-5957-0_2

MoE	Ministry of Environment
MOTAS	Motor-car Total information Advanced System
OECD	Manual of Organisation for Economic Co-operation and Development
PHEV	plug-in hybrid electric vehicles

1 Introduction

In line with China's economic sgrowth, the country's automotive industry has expanded, reaching record highs in both production and sales in 2021 and remaining the largest in the world for the thirteenth consecutive year. However, the number of end-of-life vehicles is rapidly increasing due to the growth of motorization, and there is an urgent need to establish an effective disposal and recycling system. In developed countries, recycling systems based on Extended Producer Responsibility (EPR) have been in place in a number of industries since the 1990s, and the governments of developing countries have recently been following the lead of developed countries in establishing legislation related to recycling. China is increasingly applying EPR principles to its automotive industry based on the experience of more developed countries. However, EPR-related institutional design is complex, and there are many difficulties involved in establishing a system as corporate recycling technology and public environmental awareness are not on par with the developed world. We believe that China should introduce an EPR system to its automobile industry and promote recycling to realize a more recycling-oriented society. This paper will introduce the current status of China's automotive recycling system and the current challenges of applying EPR to its automotive industry.

There have been many previous studies on China's automobile recycling systems. For example, Hiraiwa (2013) described the evolution of China's automobile recycling policy. Wang et al. (2007) and identified and clarified issues in the Chinese automobile recycling business based on comparisons between Japan and China.

The Guidance Manual of Organisation for Economic Co-operation and Development (OECD) defines EPR as "an environmental policy approach in which a producer's responsibility for a product is extended to the post-consumer stage of a product's life cycle." (OECD 2001). The Manual explains the need for the application of EPR: "The important implications and changes associated with EPR stem from both the product treatment at the post-consumer phase, addressing the upstream activities in the selection of materials and the design of the product. It is believed that, under these conditions, appropriate signals can be sent to the producer to internalize a substantial portion of the environmental externalities from the final disposal of the product." When the producer and recycler are separate economic entities, "ease-of-recycling" is not at the forefront of the producer's mind. However, this problem can be eliminated when the producer and recycler are the same entity.

This paper first discusses how the number of end-of-life vehicles is increasing due to growing motorization in China, and how the problem of automobile recycling is becoming more apparent. Then, by comparing the automobile recycling systems of Japan and China we point out the problems found in China's systems. Finally, we examine the application of EPR to the Chinese automobile industry.

2 The Development of Motorization and the Emergence of ELV in China

2.1 Trends of End-of-Life Vehicles (ELV) in Japan

Full-scale motorization in Japan began in the late 1960s. The expansion of both the domestic market and exports due to economic growth led to a dramatic increase in the automobile industry's production capacity, which reached 13 million units per year in 1990. However, domestic automobile production has been on a downward trend since 1990, as the domestic market is approaching saturation and overseas production is increasing. Figure 2.1 shows the number of automobiles owned and used in Japan. Although the number of owned vehicles continues to increase, the rate of growth has slowed since 1997, when it exceeded 70 million units. The number of automobiles per 1000 people grew from 14 in 1960 to 168 in 1970, 467 in 1990 and 572 in 2000, but there has been little change since the turn of the millennium.[1]

Conversely, while the number of end-of-life vehicles can fluctuate widely, it remains at approximately 3 to four million vehicles per year. However, about 1.5 million of these vehicles are exported overseas to be sold second-hand. It is estimated that approximately 1.5 to two million end-of-life vehicles appear annually in Japan.[2]

2.2 End-of-Life Vehicle Trends in China

Prior to the adoption of the reform and open-door policy in 1978, China produced only about 150,000 automobiles per year, with the majority being trucks.

[1]The number of vehicles per 1000 people in Japan was calculated by the author based on data for the number of owned vehicles (Japan Automobile Manufacturers Association, 2016) and the size of the population (Statistics Bureau, Ministry of Internal Affairs and Communications).

[2]Ministry of Economy, Trade and Industry and Ministry of the Environment (2000) "Current Status of Automobile Recycling."

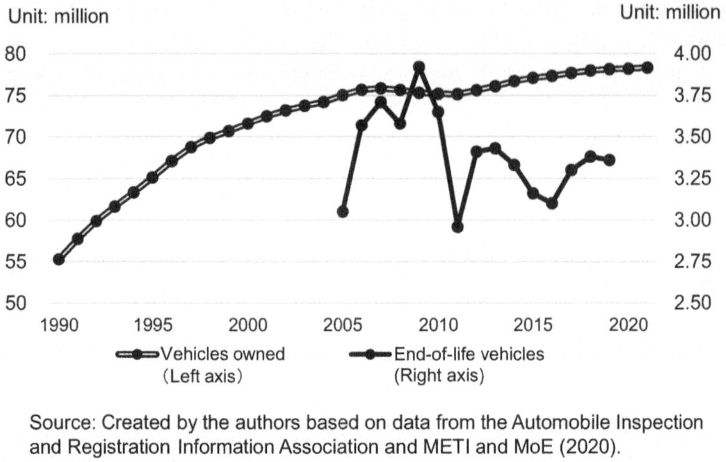

Fig. 2.1 The numbers of vehicles owned and end-of-life vehicles in Japan. Source: Created by the authors based on data from the Automobile Inspection and Registration Information Association and METI and MoE (2020)

Subsequently, the Chinese government launched a series of measures to promote the automobile industry and shifted the emphasis of automobile production from trucks to passenger cars. China's accession to the WTO in the early 2000s accelerated the growth of the Chinese economy. This increased demand for passenger cars, especially in large coastal cities, ushering in an era of full-fledged motorization. To mitigate the impact of the 2008 Lehman Brothers collapse, the Chinese government adopted "Ten Major Measures to Promote Domestic Demand and Economic Growth." In the following year, 2009, the "Automobile Industry Readjustment and Promotion Plan" was announced to stimulate automobile purchasing, and motorization expanded to inland and rural areas. In 2009, production and sales exceeded 13 million units, making it the largest automobile market in the world.

Figure 2.2 shows the number of automobiles owned and operated in China. The number of owned vehicles grew from 5.31 million in 1990 to 77.22 million in 2010, surpassing Japan's 75.36 million and ranking second-highest in the world after the U.S., which has 248.23 million. By 2014, the number of vehicles owned in China had exceeded 144.52 million and is expected to continue to increase (Japan Automobile Manufacturers Association, 2017: 105–106). The number of cars owned per 1000 people in China has increased rapidly from 5 in 1990 to 58 in 2010 and 107 in 2014,[3] in line with Japan's earlier experience of growth from 64 in 1965 to 168 in

[3]The number of cars per 1000 people in China was calculated by the author based on data on the number of cars owned (Japan Automobile Manufacturers Association 2016) and China's population (National Bureau of Statistics of the People's Republic of China 2015).

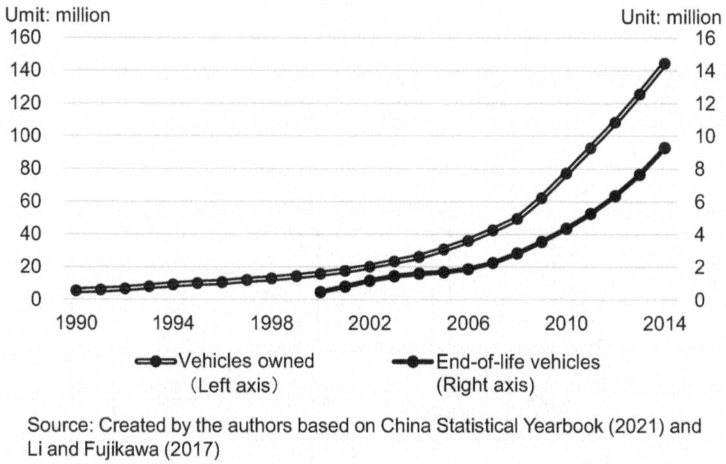

Fig. 2.2 The numbers of vehicles owned and end-of-life vehicles in China. Source: Created by the authors based on China Statistical Yearbook (2021) and Li and Fujikawa (2017)

1970.[4] China is in the early stages of motorization, which is similar to this period in Japan. In 2040, the number of automobiles owned in China is expected to reach 450 million. With a population of 1.5 billion, the number of automobiles owned per 1000 people is expected to be around 300, approximately half the current number in Japan (China Automobile Dealers Association 2015: 322).

However, the number of end-of-life vehicles generated in China was only 460,000 in 2000 but surpassed four million after 10 years, that is, 2010. The number of end-of-life automobiles in China is expected to continue to grow and is estimated to exceed 20 million vehicles by 2020 (Li and Fujikawa 2017; China Investment Consulting Network 2016).

3 Comparison of the Recycling Systems of Japan and China

3.1 The Japanese Legal Framework for Automotive Recycling

In Japan, the End-of-Life Vehicles Recycling Law, based on EPR concepts, came into effect in January 2005. The direct cause for the enactment of this law was the collapse of the existing end-of-life vehicle recycling system due to reversed charging for end-of-life vehicle disposal (Asaki 2004). Japan's automobile recycling flow is shown in Fig. 2.3. The final owner delivers the used vehicle to a salvage company

[4] As of 2014, global average for cars owned per 1000 people was 167, while the U.S. average was 809. See Japan Automobile Manufacturers Association (2016, p. 64).

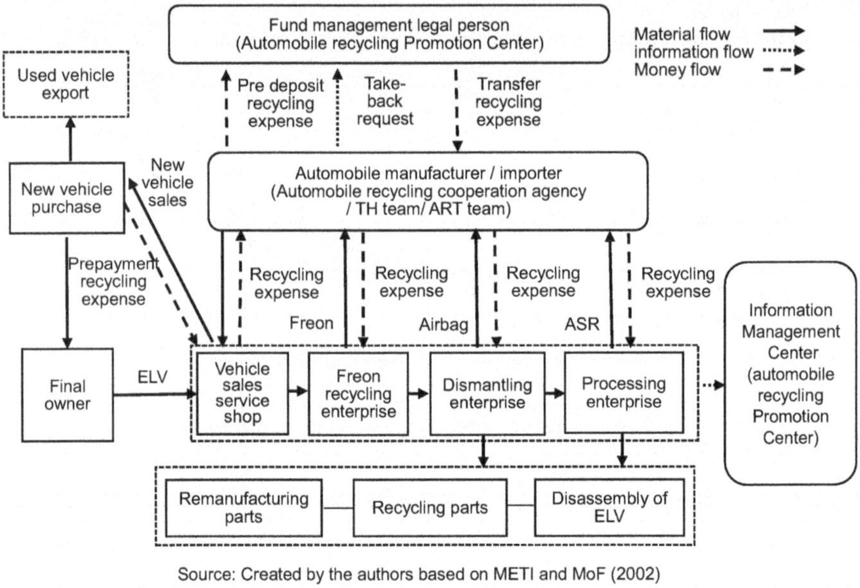

Fig. 2.3 Illustration of Japan's recycling system. Source: Created by the authors based on METI and MoF (2002)

(e.g., automobile dealers), and the salvage company delivers the used vehicle to a fluorocarbon recovery company. The recovery company recovers the fluorocarbon contained in the air conditioners of automobiles and then returns the end-of-life vehicles to their manufacturers. Under the End-of-Life Vehicles Recycling Law, automobile manufacturers are responsible for the disposal of airbags and automobile shredder residue (ASR). The automobile manufacturer delivers the used automobile to a wrecker once the fluorocarbons have been recovered. The wrecker removes the airbags, engine, doors, and other useful parts, and then delivers the dismantled vehicle to a shredder. For the recycling of airbags, automobile manufacturers have jointly established the Japan Automobile Recycling Cooperation Organization (a general incorporated association). The processing is outsourced. Shredders shred dismantled vehicles and recover useful metals. The ASR is sorted, collected, and delivered to the automaker. ASR processing is divided into two teams (the ART team and the TH team) according to the vehicle's manufacturer, and the recycling is performed by contractors from each team.[5]

Under the Japanese system, the roles of each party are clearly defined to ensure that end-of-life vehicles are disposed of properly. In the case of new cars, the buyer

[5]The ART team (Automobile Recycling Promotion Team) was formed of 13 corporations including Nissan, Isuzu, Fuji Heavy Industries, Mazda, Mitsubishi Motors, etc. The TH team (Toyota-Honda Team) was formed of 8 corporations including Toyota, Honda, Daihatsu, Hino, and others.

of a new car "prepays" the recycling fee[6] to the dealer. The fee is set by each automaker, resulting in competition among the manufacturers. The Automotive Recycling Promotion Center manages the recycling fees in case an automaker goes bankrupt. When end-of-life vehicles are exported to foreign countries to be sold second-hand, a surplus is generated. This surplus is utilized for countermeasures against illegal dumping, to subsidize remote islands, and to reduce the financial burden on vehicle owners. (Ministry of Economy, Trade and Industry and Ministry of the Environment 2002).

Japan's End-of-Life Vehicles Recycling Law has introduced an electronic manifest system. This system is a world-first in that recycling information at each stage of the process can be checked by an information management system (Japan Automobile Manufacturers Association, 2015:34). The electronic data processing system (Motor-car Total information Advanced System, MOTAS) for vehicle registration and inspection operations in Japan centrally manages registration and inspection data for vehicles owned throughout the country. A monitoring system for end-of-life vehicles tied to MOTAS has been established, and the status of vehicle recycling can be checked on the Automobile Recycling System's website.[7]

The End-of-Life Vehicles Recycling Law has promoted the reorganization of the end-of-life vehicle treatment and recycling industry. The overall recycling rate of end-of-life vehicles has increased from 83%, which was before the law came into effect, to 99% (Ministry of the Environment2015). In addition, it is also recognized as having achieved "clarification of distribution routes for end-of-life vehicles, progress in recycling by vehicle manufacturers, and a decrease in illegal dumping and improper storage"(Ueda 2010:82). At the same time, unlike the EU End-of-Life Vehicle Directive, which places physical and economic responsibility on the manufacturer, Japan's "Automobile Recycling Law" makes the automaker responsible for the disposal of fluorocarbons, airbags, and ASR through the utilization of existing infrastructure (wreckers and shredders). This method is called "Japanese EPR" and is regarded as "an efficient method tailored to Japan's social conditions"(Asaki 2004:81–82, Otsuka 2002:195, Lee 2007:324, Wang 2014, 2015:1299).

---

[6] In addition to the recycling fees for the three items (CFCs, airbags, and ASR), the charges also cover the operating costs of the Information Management Center and the Automotive Recycling Promotion Center.

[7] Automotive Recycling System http://www.jars.gr.jp/>. Funazaki (2009) mentioned this system as "an advanced system that is unique to Japan."

3.2 The Legal and Regulatory Framework for Automobile Recycling in China

Legislation related to automobile recycling in China was initiated in the 1980s. The "Truck Recovery Law" (trialed), which came into effect in 1980, stipulated procedures for the recovery of trucks and clearly stated that end-of-life vehicles were to be recovered, dismantled, and disposed off as scrap steel. The "Regulations to Accelerate the Disposal of Older Vehicles" (provisional), which came into effect in 1986, set standards for scrapping vehicles and maximum mileage limits for each vehicle type. In the 1990s, policies were implemented to strengthen the monitoring systems within the automobile industry and the market and to foster incentives to replace automobiles. The Waste Motor Vehicle Recovery Law, which came into effect in 1990, stipulates the obligations of management departments and business entities for the recovery and disassembly of end-of-life vehicles, the issuance of end-of-life vehicle recovery certificates, the acceptable fluctuation range for end-of-life vehicle purchase prices, and the prohibition of reuse of the five major components of end-of-life vehicles (engines, steering, transmission, axle shafts, and frames). The "Law on Qualification and Certification of Recovery (Dismantling) Enterprises of End-of-Life Vehicles" (provisional), which came into effect in 1997, introduced a qualification system for enterprises that recover and dismantle end-of-life vehicles. A subsidy system has also been introduced to provide incentives for purchasing new vehicles. The "Law on Fixed Subsidy for Renewal of Older Vehicles" (provisional), which came into effect in 1995, defined the scope and criteria for providing subsidies for vehicle disposal.

Table 2.1 lists the efficiencies of the automotive recycling-related regulations enacted since 2000. The "Measures for the Management of Recovery of Scrapped Vehicles" (hereinafter referred to as Decree 307), which came into effect in June 2001, will be the central law of China's automobile recycling system until the law was revised in 2019. Decree 307 stipulates the supervision and management methods for the collection and dismantling of end-of-life vehicles, the conditions for establishing a dismantling company and system for their certification, the obligations of certified companies, the procedures and purchase price of end-of-life vehicles, and penalties for related illegal acts.

The Chinese government implemented the "Road Traffic Safety Law," and the "Motor Vehicle Registration Regulations" on May 1, 2004. These regulated the vehicle registration and cancellation system, the vehicle inspection system, and the compulsory vehicle scrapping system. [8] The compulsory vehicle scrapping system was amended in 2013, as shown in Table 2.2. The new disposal regulations alter the age limits for the disposal of some vehicles. The limitation on the number of years of

[8]Vehicle registration is divided into new registration, amended registration, transfer registration, suspension of registration, and cancellation of registration. (Yano Research Institute 2008, pp. 12–14).

Table 2.1 Main regulations related to automobile recycling in China since 2000

Date	Main regulations
2001/06/16	Measures for the administration of the recycling of end-of-life vehicles
2004/05/01	Road traffic safety law
2004/05/01	Motor vehicle registration regulations
2006/02/06	Technical policy of automobile product recycling
2008/03/02	Administrative measures for pilot remanufacturing of auto parts
2009/07/13	Implementation measures for replacing old vehicles with new ones
2013/05/01	Regulations on compulsory scrapping standards of motor vehicles
2013/07/04	Pilot implementation plan of "trade in" for remanufactured products
2014/09/01	Opinions on strengthening and improving motor vehicle inspection
2015/06/01	Management requirements for hazardous substances and recyclability of automobiles
2016/01/05	Technical policy for recycling of electric vehicle power battery (2015 edition)
2017/01/04	Notice on strengthening the supervision of environmental protection compliance of second-hand vehicles
2019/06/01	Measures for the administration of recycling of end-of-life motor vehicles
2019/12/17	Technical specification for scrapped motor vehicle recycling and dismantling enterprises
2020/09/01	Detailed rules for the implementation of the administrative measures for the recovery of end-of-life motor vehicles

Source: Created by the authors based on the official websites of relevant departments of the Chinese government

use for small non-operational vehicles was abolished. Medium-sized taxis were extended from 8 to 10 years; whereas buses were shortened from 15 to 13 years.

Various subsidy programs have been established to promote the replacement and recycling of automobiles. The Automobile Recycling Implementation Law enacted in 2009 specified a policy of providing subsidies for light-duty trucks and medium-duty buses in use for less than 8 years, medium and light-duty trucks in use for less than 12 years, medium-duty buses other than taxis, and "yellow-marked vehicles[9]" that are disposed of while still in use. The 2013 "Model Implementation Plan for Promoting Used Product Recycling" specifies the implementation of a recycling promotion policy that provides a subsidy of up to 2000 yuan (¥31,584)[10] to purchasers of recycled parts.

The revised Waste Motor Vehicle Recovery and Management Act (hereinafter, Decree 715) went into effect in June 2019. Decree 715 authorized the sale of the five major discarded vehicle components to remanufacturing companies under the premise of ensuring safety to adapt to the demands of the development of a circular economy. Allowing the recycling of the five major components is expected to

[9]The term "yellow-marked vehicle" refers to gasoline vehicles that do not meet China I and diesel vehicles that do not meet China III emission standards.

[10]The exchange rate used is the average for 2013 (China National Statistics Bureau 2015). The same applies to yen conversions below.

Table 2.2 Conditions to be end-of-life vehicles by vehicle type in China

Vehicle type and usage				Expiration date (years)	Mileage (10,000 km)
Passenger car	Commercial	Taxi	Small	8	60
			Middle	10	50
			Large	12	60
		Bus		13	40
	Non Commercial	Small		–	60
		Middle		20	50
		Large		20	60
Truck	Light truck			12	50
	Small, middle and large truck			15	60
	Dangerous goods transport vehicle			10	40

Source: Created by authors based on the site of Ministry of Commerce of China http://www.mofcom.gov.cn

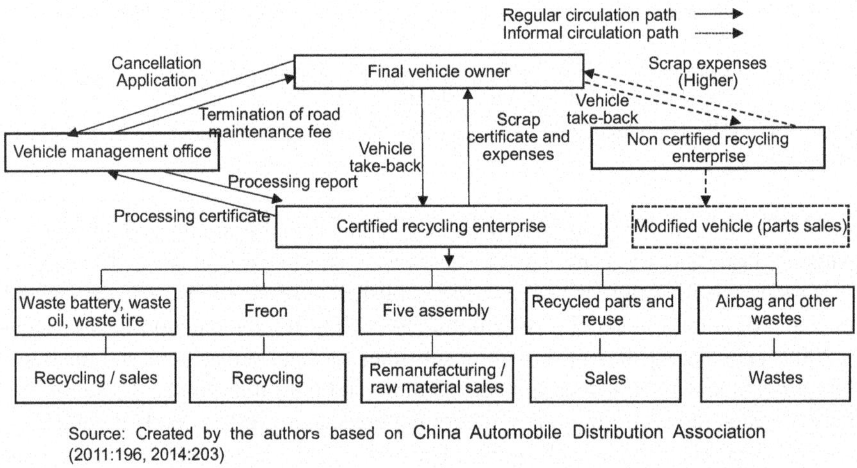

Fig. 2.4 Illustration of Japan's recycling system. Source: Created by the authors based on China Automobile Distribution Association (China Automobile Dealers Association 2011:196, 2014:203)

increase the value of recovered vehicles, thereby increasing the profits of dismantling companies and at the same time discouraging illegal vehicle disposal.

China's automobile recycling flow is shown in Fig. 2.4. End-of-life vehicles received by licensed dismantling companies (solid line) are disassembled, and fluorocarbons are recovered and airbags are disposed of as waste. This is similar to in Japan, but parts such as batteries, engines, steering wheels, transmissions, axle shafts, and frames are sold to other companies as resources. The Chinese government has enacted various laws, policies, and standards related to automobile recycling. However, its automobile recycling system does not always operate smoothly.

4 The Introduction of EPR in the Chinese Automotive Sector

The Chinese government is currently promoting the introduction of EPR. The "Automotive Product Recovery and Utilization Technology Policy" enacted in 2006 set forth the goals shown in Table 2.3 aimed at strengthening producer responsibility for automobile manufacturers and importer distributors with regard to automobile recycling. The Circular Economy Promotion Law of the People's Republic of China, enacted in 2008, was the first formal EPR legislation in China and clearly states that producers are responsible for recycling products or packaging that are on the mandatory collection list. In addition, the "Law of the People's Republic of China on the Promotion of Clean Production" revised in 2012 contains provisions on producer responsibility at the product design stage. It is said to be the "law that sets the foundation" for EPR (Wang 2015:172–173). The "Automotive Hazardous Substances and Recoverability Management Requirements" introduced in 2015 require automobile manufacturers to design vehicles to be easily recyclable and to use low-toxicity materials with a low environmental impact. This prohibited the use of six hazardous substances[11] in M1 class vehicles[12] manufactured after January 1, 2016. In addition, guidelines on the use of hazardous substances and possible recovery utilization rates are provided in the "Vehicle Production Companies and Product Public Notice," and the "Vehicle Dismantling Guidebook" is provided to recovery and dismantling companies. Furthermore, to standardize the recovery and utilization systems for electric vehicle storage batteries and also to promote the EPR system, the "Technical Policy on the Recovery and Utilization of Electric Vehicle Storage Batteries" was issued by the State Development and

Table 2.3 Three-stage target of automobile recovery / utilization technology policy

Target year	Objective	Recycle rate	Recycled material usage rate
2010	M2, M3, N2, N3 vehicle	85%	Over 80%
	M1, N1 vehicle	80%	Over 75%
2012	All domestic and imported vehicle	90%	Over 80%
2017	All domestic and imported vehicle	95%	Over 85%

Source: Created by the authors based on the official website of the government of China
Note: M1 is a passenger car with a passenger capacity of 9 or less, M2 is a passenger car with a passenger capacity of 10 or more and weighs 5 tons or less, M3 is a passenger car with a passenger capacity of 10 or more and weighs 5 tons or more, and N1 is a truck of 3.5 tons or less, N2 is a truck over 3.5 tons, and N3 is a truck over 12 tons

[11] The EU and Japanese automobile recycling laws regulate only four substances (lead, mercury, hexavalent chromium, and cadmium), but in China, six substances are regulated, including the flame retardants PBB and PBDE.

[12] From January 1, 2018, the use of hazardous substances has also been restricted for vehicles in continuous production.

Reform Commission, Ministry of Industry, Ministry of Environmental Protection, Ministry of Commerce, and General Administration of Quality Inspection on January 5, 2016. It clarified that electric vehicle and storage battery production companies are the responsible entities for the recovery and utilization of storage batteries.

5 Establishing an EV Storage Battery Recycling System

5.1 The Background of EV Introduction in China

Energy consumption in China has increased dramatically in tandem with the country's rapid economic growth. Most of the energy consumed consists of fossil fuels such as coal, petroleum, and natural gas. China notably accounts for about half of the world's coal consumption. In addition to fossil fuel consumption being a contributor to global warming, industrialization and motorization are also causing serious air pollution in urban areas. The Chinese government recognizes that future economic growth is not sustainable unless it rethinks its traditional approach. Accordingly, the introduction of renewable energy is being promoted as a means of reducing dependence on fossil fuels, restructuring energy systems, and introducing energy-efficient equipment. In the automotive field, the manufacturing and proliferation of EVs are rapidly expanding. EVs in China include battery electric vehicles (BEVs), hybrid electric vehicles, and fuel cell electric vehicles. While ordinary gasoline-powered vehicles run by burning gasoline in their engines and emit exhaust gases such as CO_2 and NO_x, EVs do not emit exhaust gases while running because the motor is powered by electricity stored in the battery. By using a motor instead of an engine, about 80% of the electric energy can be utilized for driving. In addition, the use of electricity generated from renewable energy sources such as solar and wind power can also reduce CO_2 emissions resulting from power generation.

5.2 EV Adoption in China

Figure 2.5 shows the trends in EV production in China over the period 2010–2020. EV production declined slightly in 2019 from 1,269,000 units in 2018 (BEVs: 986,000, PHEVs: 283,000), with the total number of EVs being 29,000 units less than the previous year. Meanwhile, BEV production increased to 1,020,000 units and PHEV production declined to 220,000 units. The number of BEVs and PHEVs has continued to increase since 2020, with BEVs and PHEVs up 85,000 and 125,000 units, respectively, from the previous year, making China an

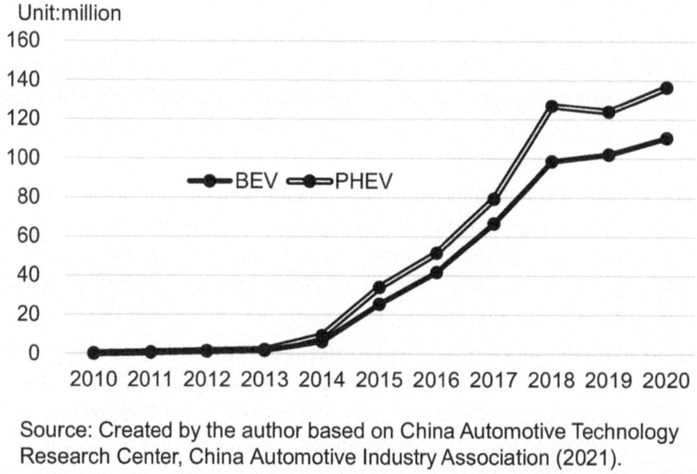

Fig. 2.5 Trends in EV production in China. Source: Created by the author based on China Automotive Technology Research Center, China Automotive Industry Association (2021)

electromagnetic vehicle powerhouse that has surpassed the United States as the producer of one-third of the world's EVs. [13]

5.3 *Recovery of EV Storage Batteries in China*

The rapid development of EVs has brought to light the problem of recovery and utilization of storage batteries, and there is an urgent need to establish a system for their collection and recycling (Li et al. 2022). Batteries for EVs include lead-acid batteries, nickel-metal hydride batteries, lithium-ion batteries, and iron phosphate ion batteries, which differ significantly from conventional batteries for electronic devices. Compared to those used in other electronic equipment, storage batteries for automobiles have a complex structure with a large weight and volume and carry a high voltage (which is itself a major safety hazard). These characteristics make reuse difficult (Winslow et al. 2018). The history of EV usage worldwide is still short, and many countries, including China, are exploring the production of EV battery recovery systems and technologies for reusing EV batteries (Chen et al. 2019). Current decommissioning companies in China lack experience in processing storage batteries due to insufficient cooperation with producers.

Currently, the manufacturer's warranty period for EV storage batteries in China is 10 years. However, battery degradation leads to a reduction in cruising range, even

[13] Prepared by the author using data from the China Automotive Technology Research Center and the China Association of Automobile Manufacturers (2021).

during the warranty period. If the remaining capacity of the battery falls below 80%, the battery cannot be used. Generally speaking, the service life of EV storage batteries is 5–10 years, so the reuse value of power storage batteries is high. Therefore, this paper examines the construction of a system for the collection and treatment of used storage batteries, primarily by EV and power battery manufacturers from the viewpoint of EPR.

Currently, production and sales in China are centered on BEVs and PHEVs, with FCVs having very low production and sales volumes. With the rapid expansion of EV manufacturing and adoption, China urgently needs to establish a system to collect and recycle used batteries. Policies such as "Technical Policy for Recovery and Utilization of Automobile Products" (Article 15) and "Technical Policy for Recovery and Utilization of EV Power Storage Batteries (2015)" promote the establishment of recycling systems for EV power storage batteries from an EPR perspective. When designing a system from the perspective of EPR, it is necessary to clarify the roles of each party involved in the recovery of EV-powered storage batteries (Hosoda 2008). The roles of key stakeholders in the recovery of EV-powered storage batteries can be summarized as follows.

EV Manufacturers and Storage Battery Manufacturers

EV manufacturers and storage battery manufacturers are responsible for collecting, recycling, and disposing of the powered storage batteries they manufacture. EV manufacturers design their EVs with the easy disassembly of the powered storage batteries in mind. The manufacturers of EVs and storage batteries collect information on the type, model, quantity, weight, distribution, etc. of their used storage batteries and report it to the regulatory authorities.

Final Owners

The final vehicle owner is obligated to hand over the used storage batteries to an authorized retrieval agent. If the EV is to be used continuously, the spent batteries must also be replaced with new ones.

Retrieval Agents

The role of the battery collection company is important because it serves as the point of contact between the final owner of the batteries and the battery disposal company. The retrieval agent collects information on the type, quantity, weight, distribution, etc., of used power storage batteries.

Battery Disassembly Processors

The role of a battery disassembly processor is to properly dismantle and recycle used batteries. The steel parts are sold to steel companies as steel raw materials, and the plastic parts are reused by waste plastic companies as recycled materials. Usable parts are reused by battery manufacturers.

Government

To promote the cascading use of storage batteries, the standardization of storage battery design and the establishment of a product code system and traceability system for storage batteries must be promoted. The government assists storage battery producers in promoting the cascading use of used storage batteries.

Based on the above, we will discuss the construction of a recovery and processing system for used power storage batteries based on current Chinese policies and the recovery systems of developed countries such as Japan from the viewpoint of EPR. The assumption for the construction of this system is that the entities responsible for the collection of EV storage batteries in China will be the EV manufacturers, the storage battery producers, and the battery cascade users. Under this assumption, a product code and traceability system for storage batteries will be established, and parties involved at each stage will register battery production, distribution, and collection and disassembly information in the "National Information Management System for Disposal and Renewal of Older Vehicles."

The issues related to the recovery of used EV batteries can be summarized in the following three points.

Normalization and Standardization of Design and Manufacture for Reuse

Storage batteries have not yet been properly codified and standardized. Although this is a difficult task for automobile manufacturers and battery producers, reusing storage batteries will be difficult without some improvement in standardization and generalization, and it is unlikely that the price of storage batteries will decrease without this.

Establish Laws Regarding the Storage and Transportation of Used Batteries

Due to the hazardous nature of used batteries during storage and transportation, specialized collectors and transporters are needed. It will also be important to establish laws regarding the storage and transportation of storage batteries in the future.

Standardization of Treatment Processes and Recycling Methods and Improvement of Treatment Technologies

The recycling of EV storage batteries, especially the treatment process and mechanism, needs to be standardized. Proper quality assurance is difficult to achieve if the batteries are handled by small auto repair shops, as the recycling of storage batteries is a labor-intensive process. Disassembly work must be centralized through the appointment of companies that have the necessary human resources and skills to carry out the work.

6 Conclusion

China's automotive industry will reach a record high in terms of both production and sales in 2021, and has been ranked first in the world for the past 13 consecutive years. However, the number of end-of-life vehicles is increasing rapidly, and there are concerns about resources wasted due to improper disposal. Therefore, there is an urgent need to establish appropriate disposal and recycling systems.

This paper reviews the changes in laws and regulations since the 1980s, and summarizes the current systems such as the eligibility certification system, vehicle deregistration system, and subsidy system. It then compares the status of the introduction of EPR in automobile industries of Japan and China. In the future, it will be necessary to examine technical issues such as the improvement of disassembly treatment technology and resource recycling technology. It will also be necessary to look at issues of awareness such as strengthening incentives for recycling-oriented design and improving the legal compliance and environmental awareness of market participants, and promoting the EPR philosophy in consideration of the social circumstances in each country. In addition, Japan was the first Asian country to introduce an EPR system, and has a better track record and technology in waste treatment and recycling than other countries in Asia. Cooperation between China and Japan in the field of recycling is an important issue that cannot be ignored, and further consideration should be given to strengthening the cooperative relationship, including the technical aspects from the Japanese side.

As an example of the application of EPR to the automotive industry, this paper examines the construction of a recycling system for storage batteries used in EVs, which have been rapidly increasing in recent years. China now faces the same environmental pollution issues due to illegal dumping and improper disposal of conventional end-of-life vehicles, just as developed countries such as Japan have experienced in the past. At the same time, it faces the problem of EV storage battery recycling, which has become an emerging issue throughout the world. In China, the introduction of EVs has been considered a key to solving environmental problems such as air pollution and CO_2 emissions caused by conventional vehicles, and the manufacturing and adoption of EVs is rapidly expanding. In China, the number of used batteries has begun to increase as EVs become more popular, and there is an urgent need to establish a collection and recycling system for these batteries.

The history of EV usage is short, and the establishment of EV battery recovery systems and recovery technologies are still being explored in China as well as in more developed countries. In China, regulatory methods such as the establishment of a product code and traceability system, and economic methods such as taxes and subsidies, have been proposed for the proper collection and disposal of storage batteries. However, the issues related to the recovery of EV used batteries can be summarized in the following three points: (1) standardization and normative design and manufacturing for the reuse of used batteries; (2) development of laws for the storage and transportation of used batteries; and (3) standardization of the treatment process and recycling method of used batteries as well as the improvement of treatment technology.

Acknowledgements This work was supported by JSPS KAKENHI Grant Numbers JP19K12459372 and JP21H04941.

References

English

Li Y, Liu Y, Chen Y, Huang S, Ju Y (2022) Estimation of end-of-life electric vehicle generation and analysis of the status and prospects of power battery recycling in China. Waste Manag Res. https://doi.org/10.1177/0734242X221080097

OECD (2001) Extended producer responsibility: a guidance manual for governments. OECD Publishing. https://doi.org/10.1787/9789264189867-en

Winslow KM, Laux SJ, Townsend TG (2018) A review on the growing concern and potential management strategies of waste lithium-ion batteries. Resources. Conservation and Recycling 129:263–277. https://doi.org/10.1016/j.resconrec.2017.11.001

Chinese

Chen J, Weng C, Lan F, Li S (2019) Development status and trend of power battery industry under the influence of policy. Science and Technology Management Research 39:148–157. (in Chinese). https://doi.org/10.3969/j.issn.1000-7695.2019.09.022

China Automobile Dealers Association (2011, 2014, 2015), China auto market almanac. China Business Press. (in Chinese)

Japanese

Asaki Y (2004) An economic analysis of the end-of-life vehicle recycling law with special reference to EPR. The Econ Rev (Kyoto University) 174(5・6):74–89. (in Japanese). https://repository.kulib.kyoto-u.ac.jp/dspace/bitstream/2433/45664/1/10174505.pdf

Funazaki A (2009) Review of end-of-life vehicle from the viewpoint of resource recycling system in Asia. JARI Res J (Japan Automobile Research Institute) 31(1):15–20. (in Japanese). https://dl.ndl.go.jp/view/download/digidepo_9215015_po_JARI696.pdf?contentNo=1&alternativeNo=&itemId=info:ndljp/pid/9215015&__lang=en

Hiraiwa Y (2013) Transition of automobile recycling policies in China : 1980s-1990s(part 2). Kogakuin Univ Bull 50(2):11–23. (in Japanese). https://iss.ndl.go.jp/books/R100000002-I023161091-00

Li Y, Fujikawa K (2017) Potential of the renewable resources of end-of-life vehicles in China. Environ Sci 30(3):184–189. (in Japanese). https://www.jstage.jst.go.jp/article/sesj/30/3/30_300303/_pdf/-char/ja. Last Access 5 Jan 2023

Otsuka T (2002) The establishment and problems of automobile recycling act. J Japan Soc Mater Cycles Waste Manag 13(4):193–199. (in Japanese). https://www.jstage.jst.go.jp/article/wmr1990/13/4/13_4_193/_pdf/-char/ja

Ueda Y (2010) Review of the automobile recycling act and future direction. J Japan Soc Mater Cycles Waste Manag 21(2):81–86. (in Japanese). https://www.jstage.jst.go.jp/article/mcwmr/21/2/21_81/_pdf/-char/ja

Wang Y (2014) On extended producer responsibility in the OECD and its introduction to Japan. J Ritsumeikan Low Sch 356:1235–1309. (in Japanese). https://www.ritsumei.ac.jp/acd/cg/law/lex/14-4/wang.pdf

Wang Y (2015) The development of extended producer responsibility in the United States and China: a comparative study with Japan. J Ritsumeikan Low Sch 359:140–202. (in Japanese). https://www.ritsumei.ac.jp/acd/cg/law/lex/15-1/004%20wang.pdf

Yano Research Institute (2008) A research report on the used car market in China. (in Japanese). http://www.econ.kyoto-u.ac.jp/~shioji/resource/Yano2008report.pdf

WEB Information

China Investment Consulting Network (2016) (in Chinese) http://www.ocn.com.cn/chanye/201607/ixgoi08143353.shtml/

METI and MoE (2002.) Outline of Automobile Recycling Law/ (in Japanese) https://www.nippo.co.jp/re_law/image/relaw8b.pdf. Last Accessed 5 Jan 2023

METI and MoE (2020) Current state of automobile recycling. (in Japanese) https://www.meti.go.jp/shingikai/sankoshin/sangyo_gijutsu/haikibutsu_recycle/jidosha_wg/pdf/048_04_00.pdf. Last Accessed 5 Jan 2023

MoE, Annual Report 2015. (in Japanese) https://www.env.go.jp/en/wpaper/2015/index.html

Chapter 3
Plastic Recycling Policy in China and the Waste Plastic Trade

Tadashi Hayashi

Contents

Keywords Plastic waste · Plastic recycling policy · Waste plastics trade · Foreign waste plastics import ban · Global recycling of plastics

[1] See New York Times (2019).

T. Hayashi (✉)
The University of Shiga Prefecture, Hikone, Shiga, Japan
e-mail: thayashi@ses.usp.ac.jp

K. Fujikawa (ed.), *Empirical Research on Environmental Policies in China*,
https://doi.org/10.1007/978-981-99-5957-0_3

1 China's Industrialization and Plastic Waste Problem

With China's rapid modernization and industrialization since the beginning of the twenty-first century, a large amount of waste has been generated, and its disposal methods and environmental problems have become major issues. Landfills are filling up faster than planned, and the amount of household waste generated is far greater than in any other country. There has been opposition to the construction of incineration facilities and landfills in many places, including in Wuhan in 2019,[1] where the waste problem could cause social unrest. "Not In My Backyard" (NIMBY) is a common reaction to new landfill development, interpreted as "I understand that we need a landfill as a community, but I am firmly against it in my own backyard." In China, where economic development continues, and waste continues to increase, this is now a major problem, and the Chinese government is struggling to deal with it.

The Chinese government has so far responded to these issues in its "Five-Year Plan," which defines the basic environmental policy direction of the government: the idea of "greatly developing the circular economy" and building a system for separate collection, recycling, and proper disposal of waste. President Xi Jinping has also pointed out the importance of "waste reduction, recycling, and detoxification" and the establishment of a waste sorting system for this purpose.[2]

China is also one of the top emitters of plastic waste such as containers and packaging.[3] The pollution of oceans and soil caused by plastic is known as *white pollution*. However, China's developing economy has increased the demand for plastic products, and the amount of plastic products disposed of has also increased. In recent years, marine plastic pollution has become a global problem, and China is said to have placed the world's largest amount of plastic in the ocean.[4]

In 2021, China restricted the importation of all overseas solid waste and the disposal of solid waste in China. In addition, the establishment of waste management systems is also being seen in various parts of China.

In this chapter, Sect. 2 looks at the Chinese government's 2017 restriction on imports of used plastics and other materials. Section 3 describes the characteristics of the circular economy in China and key laws and regulations related to waste. Section 4 summarizes the up to 2017 status and challenges of recycling policies in China, and Sect. 5 discusses the effectiveness and challenges of plastics regulations in China. The final section, Sect. 6, examines the challenges of trade in waste plastics.

---

[2] See National Committee of Chinese People's Consultative Conference (2017).

[3] See UNEP (2018) for details.

[4] See Jambeck et al. (2015) for details.

[5] See General Administration of Taxation of the People's Republic of China (various year's editions).

2 China's Ban on Waste Plastic Imports

China amended its Imported Waste Management Inventory in 2017 to restrict the import of waste plastics from household. At first glance, trade in waste plastics appears to be a revolutionary transnational recycling system, but until the implementation of the Ordinance on the Management of Household Waste in 2019, China's insufficient waste collection and disposal systems and the rapid increase in waste generation due to economic growth were seen as problematic. In fact, China's average annual plastic waste collection rate from 2009 to 2013 was 25.6%, while Japan's recovery rate during the same period was 78.4%, indicating that China's demand for raw materials imported from abroad far exceeds the supply from the domestic market.[5] In addition, new environmental problems have arisen, such as the inclusion among the waste plastics imported by China of insufficiently sorted waste plastics that cannot be turned into resources and the illegal dumping of imported waste plastics without a place to accept them.

In July 2017, the Chinese government issued the "Implementation Plan for the Reform of the Solid Waste Import Management System to Ban the Entry of Foreign Waste into the Border,"[6] which includes a ban on the importation of plastics and other solid wastes that are highly harmful to the environment and cause an "intense public outcry" by the end of 2017. Rather than banning imports of all solid wastes (or more precisely, resources for recycling), it divided the ban into two parts: an early ban by the end of 2017 on non-industrial (i.e., from household) used plastics, such as used plastic bottles, which cause a strong public reaction, and a gradual suspension of imports of other industrial derived waste plastic resources (for example, waste film plastics for logistics and packaging applications) "that can be replaced by domestic resources." It is important to note that the regulations are being implemented. The inference is that the people are against the influx of "foreign waste" in dirty domestic products and that the government is aware of this and regulates the bans more strongly than it would otherwise.

The reality of on-site recycling of used plastics imported into China was clearly illustrated in the 2016 documentary film "Plastic China." In the film, a family, including a young girl, lives among unsanitary plastic resources while the whole family works together to sort, process, and bag them, providing a major insight into various issues such as poverty, environmental pollution, and health problems. As the reality of these recycling sites came to light, the public became more and more aware of the problem, encouraging the government to adopt a policy of banning imports. In the film, the plastic "resources" include containers and packaging from Europe and the United States, as well as from Japan. This demonstrated that some of the plastic containers and packaging discarded in developed countries are turned into resources through sorting under unsanitary labor conditions, while others are simply burned. This was the reality before the import ban.

[6] See Central People's Government of the People's Republic of China (2017).

[7] See Ministry of Ecology and Environment of the People's Republic of China (2019a).

On the other hand, the government plan states that importation of resources in demand in China's domestic industries, such as resources for manufacturing, will be gradually slowed, rather than banned. This suggests that the government is cautious about completely wiping out these resources. Therefore, the government is committed to fully supplying the resources necessary to build a domestic circular economy, while also monitoring the impact of the import suspension. Therefore, the establishment of a domestic recycling system and the securing of a supply of quality resources through this system will become even more important. In fact, the Chinese implementation plan also states that the domestic waste collection and utilization rate will be raised, and that the amount of waste collected will be increased from 246 million tons in 2015 to 350 million tons by 2020. However, there are no reliable statistical data to verify this subject.

The progress of the import ban was highlighted by the Chinese government, which announced that imports of solid waste across China decreased by 46.5% y/y in 2018[7] and by 28.1% y/y in the first half of 2019, underscoring the steady progress being made.[8] China imported about 1.3 million tons of used plastic as a resource from Japan in 2017, which decreased to about 50,000 tons in 2018, indicating that the ban is being effectively enforced. The Chinese government stated that it would do its utmost to reduce the amount of imported waste to zero by the end of 2020.[9] While the import ban has been effective in this way, increasing the waste collection, and utilization rate in China order to enhance resource recycling is also developing.

3 Characteristics of the Circular Economy and Major Laws and Regulations Related to Waste in China

3.1 Characteristics of the Circular Economy in China

The term *circular economy* is an abbreviation for "closing material cycle," which first appeared officially in the *Circular Economy and Waste Management Act* promulgated by Germany in 1996. In 2000, Japan issued the *Basic Law for Establishing a Recycling-Oriented Society*, using the expression "circular-based society." Other terms related to these two concepts are mainly found in the industrial sector. These include cleaner production, ecological industry, industrial symbiosis, zero emissions, and waste minimization. There was a transition process in which the concept of a circular economy emerged as the international community shifted from the sustainable development of industry to the sustainable development of the economy and society. A circular economy is a kind of ecological economic theory and practice model. It is not a theory and practice of traditional economics on issues related to economic action. Different countries are at different socioeconomic stages

[8] See Ministry of Ecology and Environment of the People's Republic of China (2019b).

[9] See Ministry of Ecology and Environment of the People's Republic of China (2019c).

and do not share the same environmental and sustainable development issues. Therefore, the perception, and practice of the circular economy in China differ greatly from those in Germany and Japan. Thus, a circular economy, concept, and practice were formed with Chinese characteristics.

Overall, the circular economy perception with Chinese characteristics is mainly manifested in two aspects. First is its background. After developed countries gradually eliminated industrial pollution and some pollution caused by everyday life, the large amount of waste due to post-industrialization and consumer-oriented social structures has gradually become an important issue affecting environmental protection and sustainable development. Against this social and economic backdrop, a circular economy theory, and practice emerged, featuring eco-efficiency, and the conversion of waste into raw materials, recycling, and reuse (3R) as its cornerstones. China has developed its own circular economy theory and practice by referring to international experience in the process of compressed industrialization (i.e., industrial development with a dramatic transformation of the industrial structure which compressed a process into just one generation that took several generations in developed countries.) and urbanization at a low stage of development, and in the search for comprehensive strategic measures to solve complex ecological and environmental problems.

The second aspect is the content. The circular economy in developed countries started by solving the waste problem in the consumption area and then expanded to the production area. The ultimate goal was to change the "mass production, mass consumption, mass disposal" model of social and economic development. In contrast, China's current understanding and practice of the circular economy is high resource consumption and pollutant emissions and low efficiency. The intention is to change the conventional economic growth model, to balance resources, and the environment with economic development, to solve complex environmental pollution problems, to ensure the full realization of a low-cost society, and to open a new path for industrialization. Therefore, in the current implementation, industrial development and model restructuring based on cleaner production and the construction of eco-industrial areas have been given high priority.

3.2 *Major Laws and Regulations Related to Waste in China*

In China, the following three basic laws concerning waste in general have been enacted (see Table 3.1).

Solid Waste Pollution Prevention and Control Law. *This law, enacted in 1995, established rules for the disposal of industrial, domestic, and hazardous wastes, and its 2004 amendment clarified the obligation of product producers to prevent pollution generated by product waste. The 2020 amendment also clarified the principles of reduction, recycling, and detoxification in the prevention of solid waste pollution.*

Clean Produce Promotion Act. *Enacted in 2002, this law requires companies to adopt production processes that emit fewer pollutants. It also requires a production*

Table 3.1 Major laws and regulations related to waste in China

Name of Laws and Policies	Date of Promulgation, Effective Date, etc.
Solid waste pollution prevention and control Law	Passed in 1995 Enacted on April 1, 1996 April 29, 2020, the latest correction, enforced as of September 1 of the same yea
Cleaner production promotion Law	Passed in 2002 Enacted on January 1, 2003 Amended on February 29, 2012 Enacted July 1 of the same year
Circular economy promotion Law	Passed in 2008 Enacted January 1, 2009
National Hazardous Waste List	Corrected version announced June 14, 2016 Enacted august 1, 2016. Amended on November 27, 2020 Enacted January 1, 2021

Source: Prepared by the author based on various Chinese laws

process that makes it is easy to collect, recycle, and reuse materials, and packaging that is easy to collect after use.

Circular Economy Promotion Law. *Enacted in 2008, this law stipulates extended producer responsibility for waste recycling, including the integrated use of industrial waste, reuse, and* recycling *of recycled resources.*

National Hazardous Waste List. *In accordance with the "Solid Waste Pollution and Environmental Prevention Law," the National Hazardous Waste List (2021 Edition) was established. Hazardous characteristics such as toxicity, corrosiveness, flammability, reactivity, and infectivity are determined through identification based on the national criteria and methods for identification of hazardous waste.*

In addition, under the Basic Laws mentioned above, individual recycling regulations are being developed especially for automotive and electronic waste. These industries are expected to increase in volume and to yield significant economic benefits from recycling.

3.3 *Policy on Plastic Waste in China*

There is currently a movement to reduce plastic waste in countries and companies around the world. In Japan, for example, many regions have begun to charge for plastic shopping bags. The same interest applies to China.

In 2017, the Chinese government issued the "Overseas Waste Entry Ban: Implementation Plan for Reform of Solid Waste Import Management System." By 2018, this law bans the importation of plastics and other solid wastes that are highly harmful to the environment. While these measures will have a significant impact on countries that have traditionally exported waste plastics to China, some Chinese companies have invested in factories in Southeast Asia, Japan, South Korea, Europe, the United States, and other countries, therefore, benefiting these countries with China's experience and technology in collecting and processing waste plastics. While China has processed large quantities of imported waste plastics from around the world, it has never exported waste plastics to other countries and has a 100% processing rate on the mainland, a track record that has contributed significantly to waste plastic pollution control.

In 2021, the "Action Plan for Improving Plastic Pollution" was released, which states the goal of reducing plastic waste by 2025. The plan also includes efforts to strengthen the management of plastic products at every stage of production, distribution, and consumption. The plan calls for a reduction in single-use plastic products in key sectors such as retail, e-commerce, food, and beverages, home delivery, and hotels. There is an additional ban on secondary packaging of delivered goods and an increased use of renewable products.

The plan includes the steady promotion of plastic alternatives, the establishment of rules for plastic waste collection, the improvement of detoxification treatment capacity, and the strengthening of recycling. In recent years, China has collected, and utilized nearly 19 million tons of various types of plastics annually. While efficiency gains have been promoted with regard to the processing of waste plastics, issues remain with efficiency in the collection process, which must be further improved in the future.

4 Current Status and Issues of Recycling Policy in China

4.1 Waste Management in China

China produced 215,209 thousand tons of urban waste emissions in 2017, placing it second in the world's municipal waste emissions by country.[10]

According to the *China Urban Construction Statistics Yearbook*, 228.01 million tons of domestic waste was generated in 2018, an increase of approximately 5.6% over the previous year. With China's rapid population growth and economic development, the amount of domestic waste generated has been steadily increasing since 2010. In China, landfill disposal has been the main method of waste disposal, but an increasing number of cities are switching to incineration. However, China's waste treatment methods are still low tech, and landfill treatment facilities still account for more than the half of treatment facilities that render waste harmless.[11]

One of the reasons for China's low waste disposal incineration rate is the lack of waste separation. There is a past history of waste segregation efforts in Shanghai and

[10] See the web site of OECD data (2023).

[11] See Ministry of Housing and Urban Development of the People's Republic of China (Various years editions).

other cities, but they all ended in failure. Therefore, in March 2017, China announced the "Implementation Plan for the System of Separation of Household Waste" and the compulsory separation of household waste by 2020 in 46 key cities, including Beijing, and Shanghai. In July 2019, the Ordinance on the Management of Household Waste was issued in Shanghai, and "compulsory sorting" was implemented.

4.2 *Achievements and Challenges of the Ordinance on Household Waste Management in Shanghai*

In March 2017, the National Development and Reform Commission and the Ministry of Housing and Urban Construction released the "Implementation Plan for the System of Separation of Household Waste." The decision was made to preemptively implement compulsory separation (or mandatory separation with penalties) of household waste by 2020 in 46 key cities such as Beijing and Shanghai.[12]

In response, mandatory sorting has already been implemented, in one city after another, such as Qingdao City in Shandong Province, and mandatory sorting is being implemented in many parts of China. Here we will look at the achievements and challenges of the household waste management ordinance in Shanghai.

In July 2019, the Shanghai Municipal Ordinance on Household Waste Management was enforced and "compulsory sorting" was implemented in Shanghai.[13] The enforcement of the ordinance, which has been called "the strictest in history" in a major city, drew attention both domestically, and internationally.

The ordinance requires household garbage be separated into four categories: "recyclable resources" (cans, bottles, clothing, plastics, etc.), "hazardous waste" (batteries, fluorescent tubes, etc.), "wet waste" (food scraps, etc.), and "dry waste" (soiled paper waste, diapers, etc., and other garbage). The garbage is to be disposed of "at the designated time and place." The most serious penalty is a fine of up to 200 RMB (about $30) for individuals and 50,000 RMB (about $7400) for businesses that violate this ordinance. In addition, a fine of up to RMB 5000 (approximately $740) shall be imposed for willingly providing disposable daily necessities for hotels (such as toothbrushes, combs, razors, etc.) and disposable tableware and straws for food delivery service businesses (which can be substituted for paper or wood.), respectively.[14]

According to the Chinese government, food waste accounts for 56% of all household waste.[15]Removing this waste from incineration or landfill disposal is

[12] See China Government Network (2017).

[13] See Standing Committee of Shanghai Municipal People's Congress (2019).

[14] See Jiefang Daily (2019).

[15] See People's Network (2017).

[16] See Economic Reference Journal (2020).

expected to significantly impact waste reduction. In addition, food waste contains a large amount of moisture, which lowers the temperature of incineration, thereby reducing combustion efficiency, and necessitating the use of more auxiliary combustion materials. Processing this waste separately for recycling and other purposes can also reduce the amount of auxiliary combustion materials.

It is difficult to say that China's system of waste separation has been effectively implemented in practice, despite the fact that the system has existed for some time. It is important to steadily establish the habit of waste separation in the daily lives of citizens through educational activities, but it takes a long time for this practice to take root. Mandatory sorting with penalties, as proposed by the government may be an effective means of reinforcement. By the end of 2019, the number of penalties enforced under this ordinance was estimated to have reached 5546. Of these, 5085 were reportedly committed by businesses and 461 by individuals. Furthermore, 58.9% of all cases were attributed to failure to separate garbage.[16] However, as time has passed since the institution went into effect, both the public, both the public, and the regulatory authorities seem to have accumulated experience.

The city of Shanghai has many issues to be addressed. Shanghai's garbage is disposed of "at a designated place at a designated time" (as is the norm in Japan). Garbage cannot be thrown in at collection stations set up in certain areas called "company districts" outside of the designated time. This system has been criticized by some as inconvenient for businessmen and for the elderly who have poor mobility. The residents' committees in each district addressed these complaints on an individual basis, for example, by collecting door-to-door from elderly households, increasing the number of mobile stations, and extending the open hours.[17]

Volunteers assigned to each collection station provide explanations and guidance to residents. A rotating group of 20 to 30 volunteers are in charge of well-sorted stations, and they play an important role. Volunteers, however, do not have it easy: they work outdoors twice a day in 2–3 h shifts during open hours. It is a harsh work environment, especially during the summer months when the weather is extremely hot. The volunteer groups are heavily populated with retired senior citizens, but it is not easy to recruit them because the work is physically demanding. It has been pointed out that the thorough sorting that is done when volunteers are on duty suddenly becomes chaotic when they are not present to supervise the sorting. In fact, there have been reports of garbage being left on the ground at stations 9 after opening hours when there are no volunteers present.[18]

In modern China, IT technology is also playing an active role in the sorting and collection of garbage. When residents want to throw their garbage in the slot, they can launch an application on their smartphones and hold an authentication code to the screen reader, which opens the slot. This makes it possible to award points for weighing, even for a low amount of waste, which users can accumulate and redeem

[17] *Ibid.*

[18] *Ibid.*

[19] See People's Network (2019).

for cash, providing an incentive for separate collection. In addition, since it is possible to determine who put in the waste and at what time, it is easier to identify the person who put in the waste in cases of improper input.[19]

5 Effectiveness and Challenges of Plastics Regulations in China

In order to curb the creation of waste, the Chinese government has also launched a ban on the production and use of disposable plastic products. to. Below is an overview of the "Plastic Restriction Order" implemented by the Chinese government in 2008 and the "Plastic Ban Order" in 2020.

5.1 Plastic Restriction Order 2008

In 2008, the Chinese government was already enforcing the "Plastic Restriction Order"[20] as a countermeasure against white pollution caused by plastics. The order banned the production and sale of ultra-thin plastic bags and other plastic bags with a thickness of less than 0.025 mm that can easily become microplastics in the natural environment. The law also banned the free supply of plastic bags at supermarkets and other stores, and it institutionalized a fee-based system. For example, plastic bags are sold at supermarkets at a price of 0.1–0.4 RMB (about 1.5–6 cents) per bag, depending on the size.[21] Ten years later, Xinhuanet reported that "the amount of plastic bags used in supermarkets and stores has decreased by more than two-thirds, totaling about 1.4 million tons.[22]

In spite of these measures, plastic waste continues to increase in China. In addition to economic growth and increased consumption, the use of large amounts of plastic for packaging, bags, lunchbox containers, and tableware in food delivery and e-commerce delivery services has been on the rise in recent years. It is a factor in this trend. For example, in the home delivery industry, the use of plastic bags reached 14.7 billion in 2016.[23]

[20] See China Government Network (2007).

[21] See Chinanews.com (2008).

[22] See People's Network (2016)

[23] See China Net (2018).

[24] See National Development and Reform Commission and Ministry of Ecology and Environment of the People's Republic of China (2020)

5.2 The 2020 Plastic Ban Order

In response to this situation, the Chinese government has strengthened its enforcement measures. On January 19, 2020, the National Development and Reform Commission and the Ministry of Ecology and Environment issued "Opinions on Further Strengthening Measures against Plastic Pollution." This order has the three nodes 2020, the end of 2022 and the end of 2025, the different stages of the policy implementation requirements. The document states that by the end of 2020, the production, sale, and use of plastic products will be banned or restricted, and by the end of 2022, disposable plastics will be significantly reduced, alternatives will be promoted, and the ratio of resource use and energy recovery will be greatly increased by the end of 2022. In terms of geographical scope, all the built-up areas of cities above prefecture level and the built-up areas of counties in coastal areas will be in the restricted ranks.[24]

Specifically, the production and sale of disposable Styrofoam tableware and disposable plastic cotton swabs was banned by the end of 2020. The production of daily necessities containing microbeads (microscopic spherical plastic contained in facial cleansers and other products) was also banned by the end of the same year, as well as their sale by the end of 2022.

The ban on the production and sale of ultra-thin plastic bags (bags with a thickness of less than 0.025 mm), which has been implemented up to now, was continued. However, this time, in cities directly controlled cities such as Beijing and Shanghai, as well as in provincial capitals and other major cities, the use of non-biodegradable plastic bags (normal plastic bags) was banned in the end of 2020 in commercial facilities, supermarkets, restaurants, food delivery services, etc. In addition, the use of non-biodegradable plastic bags (regular ones) was banned in commercial establishments, supermarkets, restaurants, food delivery services (outlets), etc. In other words, all paid plastic bags larger than 0.025 mm was replaced by biodegradable plastic bags by the end of 2020. Biodegradable plastics are made from PLA (polylactic acid); the raw material for PLA is plants, including corn, sugarcane, potatoes, and beets.

Straws have become a symbol of disposable plastic, and the use of disposable non-degradable plastic straws was banned in the restaurant industry throughout China in the end of 2020. Similarly, the use of disposable, non-degradable plastic tableware will also be banned. Star hotels was banned in the end of 2020 and all hotels will be banned by the end of 2025 from willingly providing disposable plastic products. In addition, the delivery industry will be banned from using non-degradable plastic packaging by the end of 2025.

Those decisions to establish bans are undoubtedly a major step toward building a recycling-oriented economy. According to media reports, previous restrictive orders focused on plastic bags, but the new ban will cover all types of disposable plastics and the entire lifetime of a product: production, distribution, consumption, collection and use. The ban will therefore be more effective than previously. Also, the substitution of biodegradable plastics for single-use plastics is expected to be effective.

5.3 *Effectiveness and Challenges of the Plastics Ban*

The Chinese government's next major challenge in the area of environmental policy is enforcement. A uniform ban throughout the country for all applications would make it possible to guarantee compliance through on-site inspections of the upstream production sector, but this is not the case. The ban is phased in over a period of time in the targeted areas. For example, the ban on non-degradable plastic bags was implemented in major cities such as Beijing in the end of 2020, and was extended to small and medium-sized cities nationwide in the end of 2022. In addition, non-degradable bags will still be available for commercial use and export, although they will be banned in supermarkets and other retail outlets in the relevant cities. It is not easy to tell whether a bag used in a retail store is non-degradable or biodegradable, making it difficult to determine illegal activities. In the area of air pollution control, the government's environmental department has been effective in organizing an inspection unit to monitor violations at factories and other facilities in each region. However, the administrative cost of enforcement of a plastics ban is likely to be high because there are numerous entities subject to it.

Quickly establishing a production and supply system for biodegradable plastic products is another challenge. It will be interesting to see if this huge market will provide an opportunity for Japanese companies to make inroads. The ban on the production of microbeads has been implemented in Europe, the United States, and Thailand, but not yet in Japan (as of February 2023). The Japanese Ministry of the Environment's "Plastic Resource Recycling Strategy" (May 2019) only requires manufacturers to voluntarily regulate the use of microbeads. However, it should be noted that Japanese companies that sell products containing microbeads in Japan but produce them in China will also be affected by this measure because they will no longer be able to produce them in China. Therefore, the Japanese domestic market will be affected.

Finally, the January 2020 announcement of a strict ban on the production of plastic products with a short deadline of the end of the year, was a surprising policy even in command-and-control China. Finally, the strict ban on the manufacture of plastic products announced in January 2020, with a short deadline of the end of the year, was a surprising policy even for command-and-control China. However, white pollution of the oceans and soil is becoming more serious, and the EU is introducing restrictions on single-use plastics. Thus, the Chinese government has a sense of urgency in articulating a national strategy to expand the market for plastic countermeasures. This move can be evaluated as a clear statement of the national strategy to expand the market for plastics control. Such a move also has significant implications for the Japanese government's strategy making.

6 Problems of the Waste Plastics Trade

In 2021, the Basel Convention was amended to restrict the export of low-quality waste plastics due to the worsening marine pollution caused by plastics. Recycled resources have both resource potential and pollution potential. They are imported and recycled in developing countries when their resource potential is the important factor due to inexpensive labor costs. Depending on labor costs, recycled resources can be either good or bad, but if the resourcefulness aspect of recycled resources attracts attention, they will be traded.

Many developing countries, including China, have invested large amounts of recycled resources into economic development. Because natural and recycled resources are substitutable and recycled resources are cheaper, developing countries with lower labor costs, which are necessary for recycling resources, are the recipients of waste and recycled resources from developed countries.

Low-quality waste plastics are inefficiently recycled, and the total trade volume is large. This makes them an easy target for businesses that seek only profit from the weight recovered and disregard the risk of environmental pollution. However, in some countries, waste plastics are not well managed in terms of sorting and contamination with hazardous materials. Since the 1980s, China has been purchasing large quantities of low-quality waste plastics from Japan at low prices, processing them in China, and reusing them as recycled resources.[25] It is more profitable for companies in Japan and other developed countries to export waste plastics to China than to sell them to domestic recyclers. Therefore, companies that had been outsourcing processing to domestic companies are increasingly selling their waste plastics to China. The export trade in waste plastics from Japan to China is shown in Fig. 3.1.[26]

Export volumes have increased significantly since 2000, when Japan introduced the Basic Law for Establishing a Recycling-based Society, which enacted various recycling laws and awareness of the need to recycle increased. In 2005, China used 52.4% recycled resources for plastic products, which exceeded the use of virgin raw materials.

This law established individual recycling systems in Japan, but these systems are based on advanced domestic recycling technology. Therefore, if a large amount of used products and other materials flow overseas, excellent domestic recycling facilities become idle, and high-quality recycling stagnates. With an R&D subsidy from Japan's Ministry of the Environment, Teijin Co. Ltd. was the first company in the world to develop "bottle-to-bottle" recycling technology for used PET bottles, which began operations in November 2003. However, the company decided to

---

[25] See General Administration of Taxation of the People's Republic of China (various year's editions)

[26] Export volume of waste plastic is calculated based on HS code 3915 based on United Nations Statistics Division, *United Nations Commodity Trade Statistics Database*.

[27] Translated from Hayashi (2014)

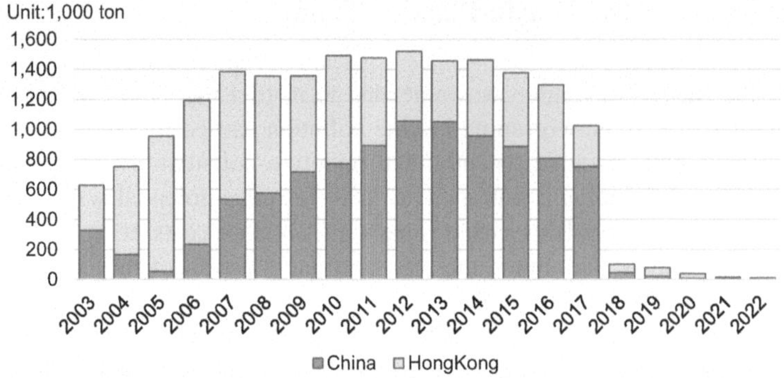

Fig. 3.1 Japan's Export Volume of Waste Plastic to China. Source: Compiled by the author. (Export volume of waste plastic is calculated based on HS code 3915 based on United Nations Commodity Trade Statistics Database)

suspend this business in 2008 due to difficulties in procuring used PET bottles in response to the rapid increase in demand for used PET bottles in China.

In Japan, the cost of recycling is relatively higher than developing countries due to permits and approvals, and there is a cost gap between domestic and foreign recycling costs with developing countries, which is also a factor in the trade of waste and recycled resources. When recycling is processed at inadequately equipped recycling plants in developing countries, there is a large risk of contamination spreading. The long chain of trade makes it extremely difficult to prevent this contamination because it is easy to lose information on the contents of traded items. As a result, information asymmetry prompted by price-cutting competition will occur, and adverse selection will proceed, with low-quality recycled resources being selected in the market.

Figure 3.2 illustrates Japan-China trade in recycled resources of waste plastics.[27] The bold arrows indicate routes with high flow rates. Once recycled resources flow into China, they are often disposed of or diffused within China. If not, plastic products recycled in China are imported to Japan; however, they cannot be recycled there because they are poor quality plastic.

Statistics on the volume of waste plastics imported by each country shows that Malaysia, Thailand, and Vietnam have received particularly large amounts of waste plastics since China's import ban. Figure 3.3 shows the evolution of waste plastics imports per month by mainland China, Malaysia, Vietnam, and Thailand from January 2016 to November 2018.[28] However, since the second half of 2018, imports to those three countries have also declined. This is because even in these three countries, import bans have begun to appear due to conspicuous environmental pollution, among other reasons. In other words, environmental policies including

[28] Compiled based on HS code 3915 (waste, parings and scrap, of plastics) from United Nations Statistics Division, *United Nations Commodity Trade Statistics Database*.

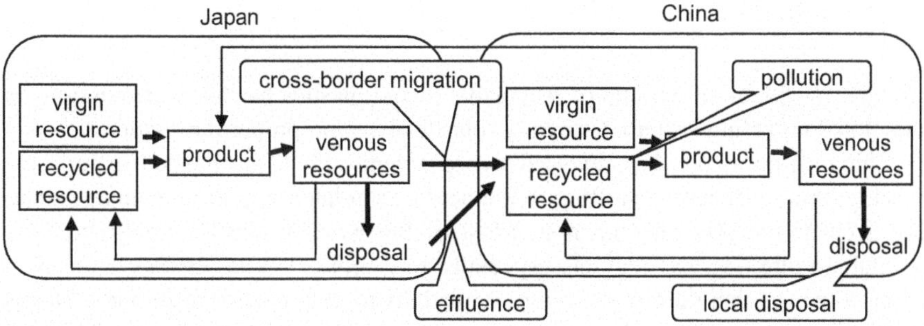

Fig. 3.2 Flow of trade in recycled resources between Japan and China. Source: Prepared by the author. Translated from Hayashi (2014)

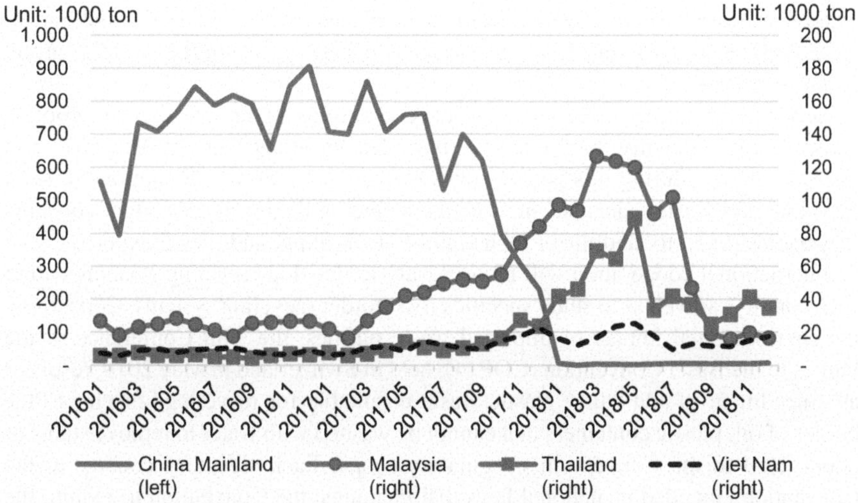

Fig. 3.3 Imports of plastic waste by China Mainland, Malaysia, Vietnam, and Thailand between January 2016 and December 2018. Source: Prepared by the author (Compiled based on HS code 3915 (waste, parings and scrap, of plastics) from United Nations Commodity Trade Statistics Database)

the import ban on waste plastics have been imposed along with the progress of economic development, the countries that accept waste plastics shifts, resulting in a chain reaction. It seems that disparities in regulations are also a source of trade. In light of the international division of labor in recycling, the international policy challenge is to coordinate cross-border policies so that there is a transborder transfer of pollution and a multi-cycle international division of labor.

7 Conclusion

China's 2018 ban on imports of recyclable mixed plastics has had repercussions in the global recycling system. Currently, plastics have no proper destination.

China's import ban has revealed two problems. First, the banned exports are quickly, but inefficiently, shifting to the next unregulated importing country, creating a chain reaction phenomenon. Most of the world's plastics flowed to less-regulated countries and regions, especially to Southeast Asia. However, even in other countries there are not enough regulations to stop excessive imports. Nor is there any real capacity to manage the waste. Malaysia, Vietnam, and Thailand became early recipients of waste plastics in anticipation of China's import ban. However, by mid-2018, less than 6 months after China's ban took full effect, these three countries had their own regulations in place. When that happened, global exports of plastic waste (mostly from the U.S., Germany, the U.K., and Japan) were simultaneously redirected to Indonesia and Turkey, two of the world's current major importers.

Second, the former exporters now have a surplus of untreated or improperly treated waste. Globally, total plastic exports fell by about half between 2016 and 2018. However, during this period, news media have reported disruptions in local recycling and waste management systems, including delays in collection, collected recyclables going to landfills or incinerators, stockpiling, and illegal exports.

International cooperation will be necessary to develop recycling systems within each country, as well as to study specific cross-border recycling systems. Concerning the environmental impact of imported waste plastics, the 14th Conference of the Parties to the Basel Convention (COP14), held in Switzerland in May 2019, resolved an amendment to add "dirty plastic waste unsuitable for recycling" (such as PET bottles of oil, plastic containers that cannot be washed with water to remove stains or odors, etc.) to the Convention's regulatory scope. Each signatory country to the Convention amended its national laws that guarantee the Convention to require the consent of the importing country before exporting waste plastics that do not meet regulations.

Prior to the revision of the Basel Convention, the EU imposed mutual notification obligations on the movement of goods in exporting, importing, and transit countries. Parties monitored trade in each other's renewable resources on a multilateral basis. Furthermore, at the June 2003 meeting of IMPEL-TFS (Implementation and Enforcement of Environmental Law - transfrontier shipment regulations of waste), it was agreed that member states would support each other in enforcing environmental regulations. The EU has a system in place whereby countries are aligned on trade in renewable resources within the EU. Therefore, the national laws, and regulations are almost identical among member countries. However, some countries provide support for countries that have difficulty implementing the content of their national laws and regulations due to differences in their economic scales. It will be necessary to establish a global recycling system, using the regime enforced in the EU as an example.

Acknowledgments This work was supported by JSPS KAKENHI Grant Number JP19K12459.

References

Central People's Government of the People's Republic of China (2017) Banning the entry of foreign garbage to promote the implementation of the reform of solid waste import management system (in Chinese). http://www.gov.cn/zhengce/content/2017-07/27/content_5213738.htm. Last Accessed 24 Jan 2023

China Government Network (2007) Notice of the General Office of the State Council on Restricting the Production, Sale and Use of Plastic Shopping Bags (in Chinese). http://www.gov.cn/zhuanti/2015-06/13/content_2879030.htm. Last Accessed 24 Jan 2023

China Government Network (2017) Notice of the General Office of the State Council on Forwarding the Implementation Plan of the National Development and Reform Commission, Ministry of Housing and Urban-Rural Development for the Domestic Waste Classification System (in Chinese). http://www.gov.cn/zhengce/content/2017-03/30/content_5182124.htm. Last Accessed 24 Jan 2023

Chinanews.com (2008) The new version of the plastic bag price of $ 0.1–0.4 each has been ready to market (in Chinese). https://www.chinanews.com.cn/cj/xfsh/news/2008/05-13/1247833.shtml. Last Accessed 24 Jan 2023

China Net (2018) Further Promotion of 'Plastic Bag Restriction Order' Requires Action by Entire Society (in Japanese). http://japanese.china.org.cn/life/2018-06/11/content_51982105_0.htm. Last Accessed 24 Jan 2023

Economic Reference Journal (2020) Waste separation tests grassroots governance (in Chinese) http://dz.jjckb.cn/www/pages/webpage2009/html/2020-01/09/content_60490.htm. Last Accessed 24 Jan 2023

General Administration of Taxation of the People's Republic of China (various year's editions), *China Customs Statistical Yearbook*

Hayashi T (2014) Trade in renewable resources and pollution exports. Research Institute for Humanity and Nature Tenchijin 23:8–9. (in Japanese)

Jambeck JR, Geyer R, Wilcox C, Siegler TR, Perryman M, Andrady A, Narayan R, Law KL (2015) Plastic waste inputs from land into the ocean. Science 347:768–771

Daily J (2019) Shanghai people pay attention! From July 1 this year, individuals mixed garbage up to 200 yuan fine (in Chinese). https://baijiahao.baidu.com/s?id=1624194207512518844. Last Accessed 24 Jan 2023

Ministry of Ecology and Environment of the People's Republic of China (2019a) Report of the State Council on the Accomplishment of the Environmental Situation and Environmental Protection Targets in 2018 (in Chinese). http://www.mee.gov.cn/ywgz/zcghtjdd/sthjghjh/201909/t20190923_748251.shtml. Last Accessed 24 Jan 2023

Ministry of Ecology and Environment of the People's Republic of China (2019b) The Second Plenary Meeting of the Inter-Ministerial Coordination Group on Banning the Entry of Foreign Waste and Promoting the Reform of Solid Waste Import Management System (in Chinese). http://www.mee.gov.cn/xxgk2018/xxgk/xxgk15/201908/t20190817_729148.html. Last Accessed 24 Jan 2023

Ministry of Ecology and Environment of the People's Republic of China (2019c) Transcript of the March 2019 Routine Press Conference of the Ministry of Ecology and Environment (in Chinese) http://www.mee.gov.cn/xxgk2018/xxgk/xxgk15/201903/t20190329_697819.html. Last Accessed 24 Jan 2023

Ministry of Housing and Urban Development of the People's Republic of China (Various years editions), *China Urban Construction Statistics Yearbook*

National Committee of Chinese People's Consultative Conference (2017) Fortnightly meeting of the CPPCC: fight a "garbage harmless treatment" lasting war for all (in Chinese). http://www.cppcc.gov.cn/zxww/2017/05/12/ARTI1494552950501778.shtml. Last Accessed 24 Jan 2023

National Development and Reform Commission and Ministry of Ecology and Environment of the People's Republic of China (2020) National Development and Reform Commission Ministry of Ecology and Environment Opinions on Further Strengthening Plastic Pollution Control (in Chinese). http://www.gov.cn/zhengce/zhengceku/2020-01/20/content_5470895.htm. Last Accessed 24 Jan 2023

New York Times (2019) Protests Over Incinerator Rattle Officials in Chinese City. https://www.nytimes.com/2019/07/05/world/asia/wuhan-china-protests.html. Last Accessed 24 Jan 2023

OECD Data (2023) Municipal Waste. https://data.oecd.org/waste/municipal-waste.htm. Last Accessed 10 June 2023

People's Network (2016) 1.4 million tons less plastic bags used in 7 years (Beautiful China - Survey) (in Chinese). http://scitech.people.com.cn/n1/2016/0217/c1007-28130400.html. Last Accessed 24 Jan 2023

People's Network (2017) People's time comment: classification, garbage can really produce value (in Chinese). http://theory.people.com.cn/n1/2017/0407/c40531-29194914.html. Last Accessed 24 Jan 2023

People's Network (2019) Beijing Municipal Regulations on Domestic Waste Management (in Chinese). http://env.people.com.cn/n1/2019/1218/c1010-31511322.html. Last Accessed 24 Jan 2023

Standing Committee of Shanghai Municipal People's Congress (2019) *Shanghai Municipal Household Waste Management Regulations* published in full, effective July, hotels shall not actively provide disposable daily necessities in guest rooms (in Chinese). http://www.spcsc.sh.cn/n1939/n1948/n1949/n2431/u1ai186702.html. Last Accessed 24 Jan 2023

UNEP (2018) Single-use plastics: a roadmap for sustainability. https://www.unep.org/ietc/ja/node/53. Last Accessed 10 June 2023

United Nations Statistics Division. United Nations Commodity Trade Statistics Database. https://library.r.chuo-u.ac.jp/mirror_db/db_UNComtrade.html. Last Accessed 24 Jan 2023

Part II
Industrial Structure and CO_2 Emissions in China

Chapter 4
Change of Industrial Structure and CO_2 Emissions in China

Zuoyi Ye and Kiyoshi Fujikawa

Contents

Keywords Structural decomposition analysis (SDA) · DPG analysis · CO_2 emissions

Abbreviation

CEDS Carbon Emission Accounts & Datasets
COP Conference of the Parties

Z. Ye (✉)
School of International Business, Shanghai International of Business and Economics, Shanghai, China
e-mail: yzuoyi@suibe.edu.cn

K. Fujikawa
Faculty of Economics, Aichi Gakuin University, Nisshin, Aichi, Japan
e-mail: fujikawa@dpc.agu.ac.jp

K. Fujikawa (ed.), *Empirical Research on Environmental Policies in China*,
https://doi.org/10.1007/978-981-99-5957-0_4

DPG deviation from proportional growth
GHG greenhouse gas
NDCs Nationally Determined Contributions
SDA structural decomposition analysis

1 Introduction

One of the most urgent challenges facing humanity today is global warming, the leading cause of which is the consumption of fossil fuels in our daily lives and production activities. At the 21st session of the Conference of the Parties (COP21) to the United Nations Framework Convention on Climate Change in Paris in 2015, it was proposed that participating countries should reduce greenhouse gas (GHG) emissions to limit the temperature rise less than 2 °C, preferably to 1.5 °C by the end of this century. Participating countries announced their own voluntary goals, NDCs (Nationally Determined Contributions), and mutually recognized these as the "Paris Agreement" to achieve this goal; however, the World Meteorological Organization (2022) estimates that the post-industrial temperature rise will reach 1.5 °C between 2022 and 2026. To prevent global warming, it is necessary to urgently promote efforts toward a low-carbon society in both developed and developing countries.

China's economy has developed rapidly, bringing a swift upsurge in energy consumption. China's carbon dioxide (CO_2) emissions surpassed the United States in 2006 to become the world's largest; as of 2021, China's CO_2 emissions are twice those of the United States. Consequently, the Chinese government, formerly inactive regarding curbing energy consumption, has changed its policy in recent years. The NDC submitted by China to the COP21 Paris conference sets a goal of reducing CO_2 emissions by 2030 and reducing CO_2 emissions per GDP by 60–65% of the 2005 levels. At the United Nations General Assembly on September 22, 2020, China's President Xi Jinping announced that China would strive to achieve carbon neutrality by 2060.

Global warming is closely related to energy consumption structure, and planning effective global warming countermeasures requires understanding the relationship between the economic structure of both supply and demand, energy consumption, and CO_2 emissions. This study clarifies the factors behind recent changes in China's industrial structure and examines how these changes are related to changes in energy demand and CO_2 emissions. For this purpose, this research uses input–output tables.

2 Literature Review

Input–output tables are economic statistics that comprehensively record the flow of goods and services. Reading an input–output table in the row direction indicates the demand structure; a certain industry's output is demanded by what industries as intermediate goods and in what form the final demand (consumption, investment, or export). Conversely, reading an input–output table in the column direction indicates the cost structure: which industries supply intermediate goods and what kind of added value (wages, profits, or indirect taxes) is generated for the output of a specific industry. By incorporating data, such as natural resource inputs or environmental pollutant emissions, input–output tables become powerful analytical tools for understanding the relationship between economic activities and resources or the environment.

Some studies use input–output tables to analyze the factors behind changes in industrial structure, energy consumption, and CO_2 emissions. Many of these studies used structural decomposition analysis (SDA), which decomposes intertemporal changes in the output of each industry into several factors. There are two types of SDA: (1) one that directly analyzes changes in the output level and (2) one that analyzes the extent of deviation after assuming that the output of each industry changes at the same rate. This study focuses on China, which has proliferated; in the case of such an economy, if we use the first method where changes in the level are analyzed, all factors would be positive, making it challenging to identify the main factors. Therefore, this study uses the second method that adjusts the scale of the intertemporal economy. This method would more appropriately be called a factor analysis of economic structural changes rather than a factor analysis of output changes.

Research on factor analysis concerning this kind of economic structural change began with the deviation from proportional growth analysis proposed by Chenery (1960). Chenery (1960) used the gross domestic product (GDP) growth rate as the proportional growth rate, defining growth industries as those with faster growth rates than the proportional growth rate and industries with slower growth rates as declining. Several demand factors explain the deviation from proportional growth defined in this way. This method is called the deviation from proportional growth (DPG) analysis. Although Chenery (1960) did not use an input–output table, this research method became an applied field of input–output analysis and produced many successors. Chenery et al. (1962) analyzed changes in Japan's economic structure from 1914 to 1954, concluding that the growth factors of growth industries were technological change (increase in intermediate demand) and import substitution. Fujikawa (1996) conducted a long-term analysis of the Japanese economy from 1914 to 1990, showing that exports replaced investment as the driving force behind changes in the industrial structure after 1970 and that a service economy exists in both intermediate and final demands in Japan. Fujikawa (1997) also analyzed the structure change in South Korea from 1960 to 1990, stating that exports drove changes in the industrial structure from the beginning of development.

SDAs targeting the Chinese economy in the early stages of the reform and opening-up policy concluded that the main factors in China's industrial structural change were import substitution, exports, and technological change (Fujikawa and Ninomiya 1997; Teng 1997; Chen and Guo 2000; Jin and Hasebe 2006). Newer studies have found that exports, investment, and technological change contributed to the change in industrial structure (Jiang 2008; Li 2015; Ye et al. 2016); however, the evaluation of private consumption is divided. For example, Jiang (2008) indicates that private consumption is a factor in the expansion, but Li (2015) and Ye et al. (2016) suggest otherwise.

Concerning the SDA on China's CO_2 emissions, Hasebe (1995) focused on 1987 to 1990, arguing that although the effects of private investment and private consumption were minor, the positive factors of intermediate inputs and exports were significant, resulting in significant increases in CO_2 emissions from the Chinese economy as a whole. Fujikawa (1997) focused on 1981 to 1992 and found that although the economy's overall CO_2 emissions increased, the first half of the study period showed a negative effect of intermediate inputs, while the second half indicated that private consumption and private investment had a negative effect. Peters et al. (2007) focused on China from 1992 to 2002, and Guan et al. (2009) focused on the period from 2002 to 2005. These studies show that while CO_2 emissions increase mainly due to urbanization (infrastructure construction) and expanding consumption in urban areas due to lifestyle changes, reductions due to technological changes and improvements in energy efficiency were marginal. Zhang (2010) focused on changes in industrial structure and CO_2 emissions from 1992 to 2002, attributing the increase in CO_2 emissions to the manufacturing sector's rapid development. Peng et al. (2015) conducted a factor analysis of changes in CO_2 emissions from 1995 to 2009, finding that CO_2 emissions increased due to the expansion of exports and final domestic demand; however, the CO_2 emissions reduction effect due to technological change was negligible. Liao and Xu (2017) conducted a factor analysis of CO_2 emissions from 2007 to 2012, finding that CO_2 emissions increased due to the expansion of final demand, and the reduction effect due to improvements in energy efficiency and changes in the input structure was not significant.

These studies concluded that technological change and improvement in energy efficiency had little effect on changes in CO_2 emissions in China, but the most recent situation in China was not reflected. After the 11th Five-Year Plan for National Economic and Social Development covering 2006–2010, the Chinese government started full-scale efforts to improve the utilization efficiency of resources and energy and protect the environment. Therefore, this paper estimates the connected input–output table for 2007–2012 – 2017 and follows Fujikawa's method (1997) to conduct an SDA on changes in China's industrial structure and CO_2 emissions.

3 Model and Data

3.1 DPG Analysis on Output

This section introduces a model formula that applies DPG analysis, a method of assuming a hypothetical situation in which all industries have changed at the same growth rate and explaining the divergence between this and the actual inter-industry structure (this is the deviation from proportional growth). In the input–output framework, changes in production technology (changes in input coefficients), changes in final demand, and changes in the import ratio explain deviations from proportional growth.

α represents the growth ratio (proportional ratio) of the total output from the first to the second period, and the DPG of each industry is defined as follows:

$$\Delta \mathbf{x} = \mathbf{x}_2 - \alpha \mathbf{x}_1 \tag{4.1}$$

where $\mathbf{x}_1$ and $\mathbf{x}_2$ represent the output in the first and second periods, respectively. DPG analysis decomposes the DPG defined in this way into various demand factors using the input-output analysis framework. The supply and demand balance equation for the first period can be expressed as follows:

$$\mathbf{x}_1 = \left(\mathbf{I} - \widehat{\mathbf{M}}_1\right)\mathbf{A}_1\mathbf{x}_1 + \left(\mathbf{I} - \widehat{\mathbf{M}}_1\right)\mathbf{d}_1 + \mathbf{e}_1 \tag{4.2}$$

where $\widehat{\mathbf{M}}_1$ is a diagonal matrix with the import ratio and $\mathbf{A}_1$is an input coefficient matrix. $\mathbf{d}_1$, $\mathbf{e}_1$ vectors represent final domestic demand (the sum of consumption, capital investment, and the net increase in inventories) and exports, respectively. Domestic demands are converted to demands for domestically produced goods by multiplying $\left(\mathbf{I} - \widehat{\mathbf{M}}_1\right)$ from the left side; however, note that all exports are assumed to be domestically produced, so no import coefficient is multiplied. Solving Eq. (4.2) for output $\mathbf{x}_1$ yields the following equation for determining equilibrium output:

$$\mathbf{x}_1 = \left[\mathbf{I} - \left(\mathbf{I} - \widehat{\mathbf{M}}_1\right)\mathbf{A}_1\right]^{-1}\left[\left(\mathbf{I} - \widehat{\mathbf{M}}_1\right)\mathbf{d}_1 + \mathbf{e}_1\right] \tag{4.3}$$

Following the same method, the equation for determining equilibrium output can be obtained for the second period.

$$\mathbf{x}_2 = \left[\mathbf{I} - \left(\mathbf{I} - \widehat{\mathbf{M}}_2\right)\mathbf{A}_2\right]^{-1}\left[\left(\mathbf{I} - \widehat{\mathbf{M}}_2\right)\mathbf{d}_2 + \mathbf{e}_2\right] \tag{4.4}$$

Substituting Eqs. (4.3) and (4.4) into eq. (4.1), we obtain the following model equation that explains the DPG.

$$\Delta\mathbf{x} = \mathbf{B}_2\left(\mathbf{I} - \widehat{\mathbf{M}}_2\right)\Delta\mathbf{d} + \mathbf{B}_2\Delta\mathbf{e} + \mathbf{B}_2\left(\widehat{\mathbf{M}}_1 - \widehat{\mathbf{M}}_2\right)\alpha(\mathbf{A}_1\mathbf{x}_1 + \mathbf{d}_1 + \mathbf{e}_1)$$
$$+ \mathbf{B}_2\left(\mathbf{I} - \widehat{\mathbf{M}}_2\right)(\mathbf{A}_2 - \mathbf{A}_1)\alpha\mathbf{x}_1 \quad (4.5)$$

where $\mathbf{B}_2 = [\mathbf{I} - (\mathbf{I} - \mathbf{M}_2)\mathbf{A}_2]^{-1}$ expresses the Leontief inverse matrix in the second period. The first and second terms on the right-hand side are the DPG. The first results from the fact that the growth rate of final domestic demand differed from that of the total output, and the second term is the DPG resulting from the fact that the growth rate of export differed from that of the total output. The third term is the DPG resulting from changes in import coefficients (changes in import dependency), and the fourth is the DPG resulting from changes in input coefficients (technological change).

3.2 DPG Analysis on CO_2 Emissions

DPG of CO_2 emissions of each industry can be divided into production and emission coefficient (CO_2 emissions per output) factors.

$$\Delta\mathbf{c} = \mathbf{c}_2 - \boldsymbol{\alpha}\mathbf{c}_1 = \mathbf{p}_2\mathbf{x}_2 - \mathbf{p}_1\boldsymbol{\alpha}\mathbf{x}_1 = \mathbf{p}_2(\mathbf{x}_2 - \boldsymbol{\alpha}\mathbf{x}_1) + (\mathbf{p}_2 - \mathbf{p}_1)\boldsymbol{\alpha}\mathbf{x}_1 \quad (4.6)$$

where **c** and **p** stand for CO_2 emissions and emission coefficient, respectively. The first term on the right-hand side is the effect of changes in domestic output, and the second is the effect of changes in emission coefficient. $(\mathbf{x}_2 - \boldsymbol{\alpha}\mathbf{x}_1)$ in the first term on the right-hand side is the DPG of domestic output; thus, substituting Eq. (4.5) into Eq. (4.6) can decompose eq. (4.6) into the factors of changes in demand items and changes in technological structure, as shown in eq. (4.7).

$$\Delta\mathbf{c} = \mathbf{p}_2\mathbf{B}_2\left(\mathbf{I} - \widehat{\mathbf{M}}_2\right)\Delta\mathbf{d} + \mathbf{p}_2\mathbf{B}_2\Delta\mathbf{e} + \mathbf{p}_2\mathbf{B}_2\left(\widehat{\mathbf{M}}_1 - \widehat{\mathbf{M}}_2\right)\alpha(\mathbf{A}_1\mathbf{x}_1 + \mathbf{d}_1 + \mathbf{e}_1)$$
$$+ \mathbf{p}_2\mathbf{B}_2\left(\mathbf{I} - \widehat{\mathbf{M}}_2\right)(\mathbf{A}_2 - \mathbf{A}_1)\alpha\mathbf{x}_1 + (\mathbf{p}_2 - \mathbf{p}_1)\boldsymbol{\alpha}\mathbf{x}_1 \quad (4.7)$$

The DPG total of all industries is zero in the case of DPG of output, but in the case of CO_2 emission DPG, each industry's total is not zero. A positive DPG on CO_2 emissions indicates that structural changes resulted in higher CO_2 emissions relative to output. Conversely, a negative DPG on CO_2 emissions indicates that structural changes resulted in lower CO_2 emissions relative to output.

3.3 Data

The input–output tables published in China in 2007, 2012, and 2017 are nominal (National Bureau of Statistics 2009, 2015, 2019). It is necessary to create a link input–output table with fixed prices to exclude the effects of prices and compare industrial structure changes over time. The China Statistical Yearbook (National Bureau of Statistics 2008-2019) figures were used as price data, and the CO_2 emission coefficients (CO_2 emissions per output) by industry were obtained from CEADs (Carbon Emission Accounts & Datasets). The industrial classification of these statistics is not the same as the industrial classification of the input–output table; therefore, it was necessary to adjust the industrial classification among statistics. Table 4.1 shows the correspondence between the input–output table and the CO_2 emission coefficient data by industry classification. After adjusting the industrial classifications, the input–output table used in this study has 28 sectoral classifications.

We recalculated the 2007 base price index since the CSY carries a chain price index with the previous year's price set to 100. For the prices of agriculture, forestry, and fisheries, we used the CSY Agricultural Product Producer Price Index. When the mining and manufacturing sectors' industry classification of the price data in the CSY matches that in the input–output table, unaltered CSY prices are used. When the sector classification of the input–output table is rougher than that of the CSY price index, we used the weighted average of the CSY price data with the weight of the output in the input–output table. Furthermore, we used the Construction and Interior Works Price Index in CSY for the price of the construction sector and the Consumer Price Index in CSY for transportation, postal services, telecommunications, commerce, lodging, restaurants, and other services. With the industry-specific prices created above, we created a fixed price time series input–output table for 2007, 2012, and 2017, using 2007 as the base year. We also used the domestic goods' prices for imports and exports.

4 Changes in China's Industrial Structure

4.1 Growth Source of the Macroeconomy

The average growth rate in the first period of 2007–2012 (α_1) was 1.71, and that in the second period of 2012–2017 (α_2) was 1.38.

Figure 4.1 shows the results of DPG analysis on the macroeconomy. Investment, import substitution, and consumption were the main positive factors in the first half, while exports and technological change were the negative factors. In the second period, consumption, investment, and import substitution were the main positive factors, while technological change and exports were negative.

Table 4.1 Industry classification correspondence table

NO	CO_2 Emissions Database (CEADs)	NO	Input-Output table with 42 sectors	NO	Aggregated Input-Output table with 28 sectors
1	Agric, forestry & fisheries	1	Agric, forestry & fisheries	1	Agric, forestry & fisheries
2	Coal	2	Coal	2	Coal
3	Crude oil & natural gas	3	Crude oil & natural gas	3	Crude oil & natural gas
4 5	Iron ore Nonferrous mining	4	Metal mining	4	Metal mining
6 7	Non-metallic mining Other mining	5	Non-metallic mining	5	Non-metallic mining
9 10 11 12	Processed food Food processing Beverage Tobacco	6	Food &Tobacco	6	Food & Tobacco
13 24	Fiber Chemical fiber	7	Fiber	7	Fiber
14 15	Clothing Leather	8	Clothing & Leather	8	Clothing & Leather
8 16 17	Logging out Wood processing Furniture	9	Wood processing & furniture	9	Wood processing & furniture
18 19 20	Papermaking Printing Stationary	10	Papermaking, printing & stationary	10	Papermaking & printing
21	Petroleum & coal prods	11	Petroleum & coal prods	11	Petroleum & coal prods
22 23 25 26	Chemical products Pharmaceuticals Rubber product Plastic products	12	Chemical industry	12	Chemical industry
27	Ceramic & clay	13	Ceramic & clay	13	Ceramic & clay
28 29	Iren & steel Nonferrous metal	14	Primary metal	14	Primary metal
30	Metal products	15	Metal products	15	Metal products
31 32	General machinery Special machinery	16 17	General machinery Special machinery	16	General machinery
33	Transport machinery	18	Transport machinery	17	Transport machinery
34	Electric machinery 1	19	Electric machinery	18	Electric machinery
35	Electronic & comm eqp	20	Electronic & comm eqp	19	Electronic & comm eqp
36	Precision equipment	21	Precision equipment	20	Precision equipment
37 38	Other manufacturing waste disposal	22 23 24	Other manufacturing waste disposal Machine repair	21	Other manufacturing
39	Power & heat supply	25	Power & heat supply	22	Power & heat supply
40	Gas prod & supply	26	Gas prod & supply	23	Gas prod & supply

(continued)

Table 4.1 (continued)

NO	CO_2 Emissions Database (CEADs)	NO	Input-Output table with 42 sectors	NO	Aggregated Input-Output table with 28 sectors
41	Water supply	27	Water supply	24	Water supply
42	Construction	28	Construction	25	Construction
43	Transport, postal service & communication	30 32	Transport & postal service Communication	26	Trans & Comm
44	Commerce & restaurant	29 31	Commerce Hotel & restaurant	27	Commerce & restaurant
45	Other services	33 34 35 36 37 38 39 40 41 42	Finance Real estate Rental business Scientific research Public management Personal service Education Public health Culture, sports & Ent Public administration & social security	28	Other services

Source: Authors' compilation based on CEADs (Carbon Emission Accounts & Datasets) https://www.ceads.net/ and China Input–output table 200, 2012, and 2017

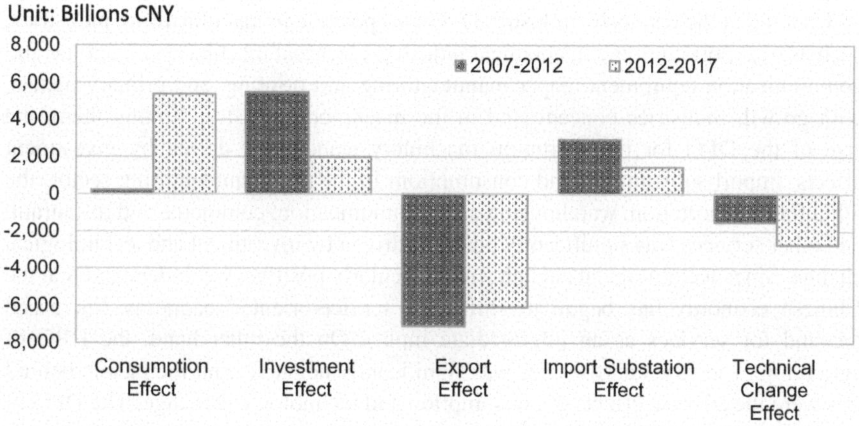

Fig. 4.1 DPG analysis on output (macro). Source: Created by authors

The positive and negative factors patterns were the same in both periods; however, the figures' magnitude changed considerably. The positive effect of consumption expanded rapidly in the second period. Although investment was a positive factor, its magnitude decreased from the first period, when the world economy experienced stagnation due to the global economic crisis originating in the United States in 2008 (known as the Lehman shock in Japan) and the European debt crisis in

2009 (known as the Euro crisis in Japan). As a result, the Chinese economy struggled to continue its export-driven economic growth.

Nonetheless, the Chinese government expanded public investment (infrastructure investment, capital investment in key industries) to support the economy [1] and adopted a low interest-rate policy, leading to a real estate investment boom. [2] Additionally, tax cuts on passenger cars started in 2015, stimulating investment and consumption in the private sector, and the expansion of Internet sales has increased personal consumption.

4.2 DPG Analysis on Output 2007–2012

Table 4.2 shows the results of the DPG analysis from 2007 to 2012 (first period). As already mentioned, China's exports during this period were sluggish due to the global financial crisis; however, the Chinese government made large-scale infrastructure investments and capital investments in key industries to stimulate the economy. Therefore, investment was the major positive factor (5464 billion CNY) for industrial structural changes during this period, followed by import substitution (2925 billion CNY) and consumption (222 billion CNY). The largest negative factor was the export effect (7101 billion CNY), followed by technological change (1510 billion CNY).

Looking at the results by industry, DPG was positive in manufacturing industries, such as transport equipment, chemical industry, electrical machinery, electronic and communication equipment, paper manufacturing and printing, and primary metals, with growth industries concentrated in the machinery industry. Among them, the size of the DPG for transportation machinery stands out, driven by investment effects, import substitution, and consumption. In the non-manufacturing sector, the DPG for transportation, warehousing and communication, commerce and restaurant, and other services was significantly positive, driven by investment and technological change. Since technological change was particularly positive, we understand that the Chinese economy has begun to shift to a service-oriented economy (increased demand for services as an intermediate input). On the other hand, the DPG of agriculture, forestry, and fisheries was significantly negative, and the factors behind this were the adverse effects of consumption and technological change. The DPG of textiles was also significantly negative, mainly due to the decrease in exports.

---

[1] The domestic demand expansion policy announced by the Chinese government is known as the four trillion CNY large-scale economic stimulus package. See Japan External Trade Organization (2009) for details. Four trillion CNY is about 600 billion USD as of 2008.

[2] Please see Sect. 2: Response to global excess production capacity, Chap. 2, Part II, of the White Paper on International Economy and Trade 2018, Ministry of Economy, Trade and Industry (2018).

Table 4.2 DPG analysis on output by industry (2007–2012) (α = 1.71; Unit: Billions CNY)

Industry		DPG	Cons	Inv	Export	Import Sub	Tech Change
1	Agric, forestry & fisheries	−2238.9	−912.8	95.3	−422.2	−27.0	−972.3
2	Coal	−194.5	16.8	56.3	−95.6	−78.9	−93.0
3	Crude oil & natural gas	−742.8	42.2	54.8	−40.0	−411.5	−388.3
4	Metal mining	−23.5	4.5	27.9	−73.2	15.3	2.0
5	Non − metallic mining	−276.9	−1.0	−1.0	−21.4	8.2	−261.6
6	Food & Tobacco	320.3	7.0	61.5	−270.9	42.2	480.6
7	Fiber	−1520.0	−71.4	19.6	−1562.6	104.3	−9.9
8	Clothing & leather	−27.0	−34.2	20.3	154.7	−22.9	−145.0
9	Wood processing & furniture	−183.9	34.3	−48.2	−95.7	−2.1	−72.1
10	Papermaking & printing	190.6	157.9	141.4	161.8	47.0	−317.4
11	Petroleum & coal prods	−962.2	109.0	151.7	−108.3	33.4	−1148.1
12	Chemical industry	489.0	148.3	196.8	−984.4	897.7	230.6
13	Ceramic & clay	102.9	−23.1	31.1	−19.8	49.3	65.4
14	Primary metal	123.4	46.1	448.9	−623.1	155.5	95.8
15	Metal products	−62.5	−6.5	196.5	−233.7	89.2	−108.2
16	General machinery	124.7	40.3	657.0	−58.5	437.6	−951.7
17	Transport machinery	711.4	319.4	1086.5	−88.6	20.5	−626.4
18	Electric machinery	361.7	46.8	179.5	−180.7	319.5	−3.5
19	Electronic & comm eqp	225.0	86.8	49.1	−589.8	499.1	179.8
20	Precision equipment	−281.3	−0.2	−28.6	−411.6	275.5	−116.4
21	Other manufacturing	−1238.2	−179.1	−96.9	−240.1	−219.3	−502.8
22	Power & heat supply	−1057.4	85.1	135.7	−233.2	71.4	−1116.4
23	Gas production & supply	59.4	54.7	13.6	−5.3	2.0	−5.5
24	Water supply	−57.6	13.1	2.1	−4.0	1.2	−70.0
25	Construction	197.8	−159.0	25.5	−25.1	27.7	328.6
26	Trans & Comm	1127.1	199.2	980.5	−366.1	38.6	274.9
27	Commerce & restaurant	465.0	−438.9	303.9	11.5	113.2	475.3
28	Other services	4368.4	636.9	702.9	−675.6	438.3	3265.9
	Total	−0.0	222.4	5463.8	−7101.4	2924.9	−1509.8

Source: Authors' calculation

4.3 *DPG Analysis on Output 2012–2017*

Table 4.3 shows the results of the DPG analysis for 2012–2017 (second period). Consumption was the largest positive factor (5382 billion CNY) for industrial structural changes during this period, followed by investment (1994 billion CNY) and import substitution (1447 billion CNY). Before the Lehman shock, China's exports were brisk, and the accompanying investment was the engine of growth; however, economic environments changed in China after the Lehman shock. The large−scale economic stimulus package of four trillion CNY and the low interest

Table 4.3 DPG analysis of output by industry (2012–2017) (α = 1.38; Unit: Billions CNY)

Industry		DPG	Cons	Inv	Export	Import Sub	Tech Change
1	Agric, forestry & fisheries	−1291.0	2.0	−528.6	−130.1	−24.6	−609.8
2	Coal	−342.7	26.0	36.3	−50.2	28.9	−383.7
3	Crude oil & natural gas	109.4	45.2	26.9	−40.2	96.8	−19.2
4	Metal mining	−350.9	10.5	−4.3	−36.3	−83.1	−237.7
5	Non − metallic mining	22.1	2.9	53.4	−8.4	−5.0	−20.7
6	Food & Tobacco	247.7	925.0	−150.0	−160.9	−103.7	−262.8
7	Fiber	−1005.1	−74.5	−62.3	−352.9	−25.6	−489.8
8	Clothing & leather	−476.2	−145.0	−4.5	−372.1	−45.6	91.0
9	Wood processing & furniture	−85.7	57.0	110.7	−63.4	−47.2	−142.7
10	Papermaking & printing	−184.6	148.0	−8.5	−217.9	35.1	−141.4
11	Petroleum & coal prods	−560.2	114.0	101.0	−83.2	46.3	−738.4
12	Chemical industry	−949.9	375.8	93.6	−582.9	−164.4	−672.1
13	Ceramic & clay	−50.3	10.4	529.9	−102.9	15.9	−503.6
14	Primary metal	−4120.8	98.9	−30.9	−348.6	166.3	−4006.5
15	Metal products	−78.5	37.4	−80.4	−145.9	45.4	65.1
16	General machinery	−1888.9	71.5	−1665.9	−367.6	34.1	38.9
17	Transport machinery	−129.0	453.1	−584.4	−356.1	87.3	271.1
18	Electric machinery	−685.2	−38.1	−465.2	−286.6	63.0	41.7
19	Electronic & comm eqp	1204.0	256.6	−76.5	−965.4	1163.7	825.6
20	Precision equipment	102.2	18.9	−47.5	−39.9	76.6	94.0
21	Other manufacturing	96.0	18.3	−8.7	−14.5	142.1	−41.2
22	Power & heat supply	−809.6	54.1	75.4	−149.3	16.1	−806.0
23	Gas production & supply	112.9	14.5	5.7	−6.0	0.3	98.5
24	Water supply	−9.4	−8.3	5.2	−2.8	0.3	−3.8
25	Construction	2003.0	10.5	2273.6	−25.2	−13.6	−242.3
26	Trans & Comm	4108.1	1001.4	1248.7	−171.2	−9.4	2038.7
27	Commerce & restaurant	1053.9	99.5	86.2	−487.5	−77.3	1432.9
28	Other services	3958.9	1796.5	1065.5	−526.3	27.8	1595.3
	Total	0.0	5382.2	1994.2	−6094.1	1446.6	−2728.8

Source: Authors' calculation

rate policy ended in 2010, after which China's economic policy shifted to a tightening one. In the latter period (2012–2017), the effects of the large−scale economic stimulus measures diminished, and the loan interest rate remained relatively high (approximately 6%). Therefore, the excess production capacity created by the economic stimulus measures became a problem. [3] Consumption replaced investment

[3]Please see Sect. 2: Response to global excess production capacity, Chap. 2, Part II, of the White Paper on International Economy and Trade 2018, Ministry of Economy, Trade and Industry (2018).

as the center of economic growth. During this period, the Chinese economy was at a turning point from export-led to domestic demand-led, and within domestic demand, from investment-led to consumption-led. Conversely, technological change (−2729 billion CNY) was the largest negative factor, followed by exports (−6094.1 billion CNY).

Looking at the results by industry, DPG was negative in most manufacturing sectors; the electronic and communication equipment sector had the only relatively large positive figure. DPG was also positive in precision equipment, food and tobacco, and other manufacturing; however, the magnitude was marginal. In contrast, China's real estate investment boom continued, and the construction industry's DPG proliferated. Other industries with positive DPGs included other services, transportation and communications, and commerce and restaurants. In addition to the consumption effect, technological change was significant in these industries, and China's economy became increasingly service-oriented.

5 Changes in China's CO_2 Emission Structure

5.1 Source of Changes in China's CO_2 Emissions

Figure 4.2 shows the results of the DPG analysis on the CO_2 emissions. During the first period (2007–2012), the economy expanded 1.7 times; simultaneously, CO_2 emissions increased no more than 1.4 times from 6.2 billion tons to 8.7 billion tons. The increase rate in CO_2 emissions is lower than that in output, indicating that energy efficiency improved. Technological change and changes in emission coefficients explain a significant part of the energy intensity improvement. CO_2 emissions

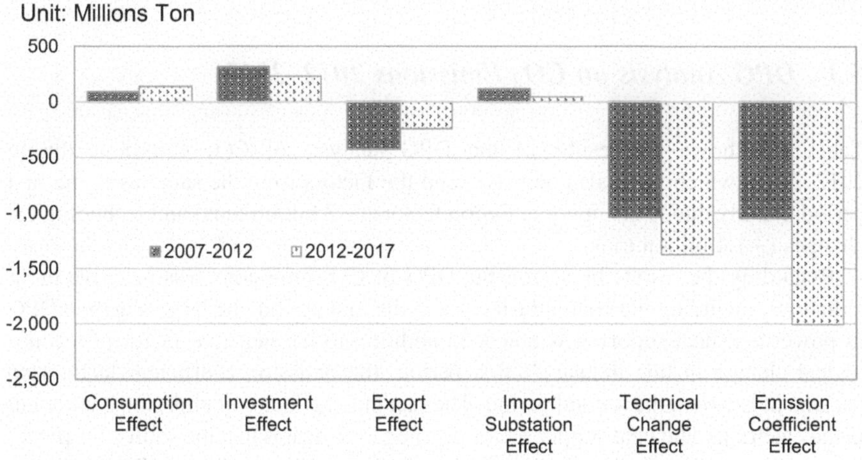

Fig. 4.2 DPG analysis on CO_2 emissions (macro). Source: Created by authors

increased due to consumption, investment, and import substitution; however, those impacts were limited. In addition, during the second period (2012–2017), while the economic scale expanded 1.4 times, CO_2 emissions increased slightly from 8.7 billion to 8.9 billion tons. As in the first half, technological change and changes in emission factors were the primary sources.

5.2 DPG Analysis on CO_2 Emissions 2007–2012

Table 4.4 shows the results of the DPG analysis on CO_2 emissions during 2007–2012; the GDPs in CO_2 emissions were mainly negative due to changing emission factors (−1 billion tons) and technological changes (−1 billion tons).

Regarding the results by industry, the DPG of CO_2 emissions was negative in all industries except coal. Notably, the reduction in CO_2 emissions from electricity and heat supply is extremely large among such industries. Technological change was the primary source of the negative DPG of electricity and heat supply, suggesting that energy conservation (especially power saving) has progressed in each industry in China. It is also noteworthy that electricity and heat supply account for most technological change factors for reducing CO_2 emissions. In the manufacturing sectors, the negative DPG of CO_2 emissions is (negatively) large in industries like primary metals, ceramic and clay, and chemical industries, and the primary source for this is the decrease in the emission coefficients. Like the previous section, in the non − manufacturing sectors, such as transportation and communications, commerce and restaurants, and other services, the DPG on output was positive due to the increase in demand; however, the decrease in the emission coefficient is large enough to offset it, resulting in the DPG in CO_2 emissions of these sectors becoming negative.

5.3 DPG Analysis on CO_2 Emissions 2012–2017

Table 4.5 shows the results of the DPG analysis of CO_2 emissions during 2012–2017, which were also negative, and the factors were the same as in the first period, mainly due to changing emission factors (−2 billion tons) and technological changes (−1.4 billion tons).

Regarding the results by sector, the DPG of CO_2 emissions was negative in all industries, including the coal industry. As in the first period, the large negative DPG of power and heat supply was noted. In addition to the negative factor of technological change in this industry's first period, the emission coefficient factor also became negative in the second period. The fact that the factor of emission coefficient in the electricity and heat supply sector was negative means that the source for power generation has become low−carbonized. The significant negative DPG of CO_2 emissions in the manufacturing sector was recorded in primary metals, ceramic

Table 4.4 DPG analysis of CO_2 emissions by Industry (2007–2012) (Unit: Millions ton)

	Industry	DPG	Cons	Inv	Export	Import Sub	Tech Change	Emi Coeff
1	Agric, forestry & fisheries	−45.5	−13.5	1.4	−6.2	−0.4	−14.4	−12.5
2	Coal	8.2	1.4	4.8	−8.2	−6.8	−8.0	24.9
3	Crude oil & natural gas	−33.0	2.0	2.6	−1.9	−19.6	−18.5	2.4
4	Metal mining	−2.3	0.1	0.4	−1.2	0.2	0.0	−2.0
5	Non−metallic mining	−8.2	−0.1	−0.1	−1.4	0.5	−17.1	9.9
6	Food & Tobacco	−64.2	0.1	0.8	−3.6	0.6	6.4	−68.5
7	Fiber	−45.7	−1.0	0.3	−21.3	1.4	−0.1	−24.9
8	Clothing & leather	−10.9	−0.1	0.1	0.5	−0.1	−0.5	−10.8
9	Wood processing & furniture	−8.5	0.3	−0.4	−0.7	−0.0	−0.5	−7.1
10	Papermaking & printing	−31.6	2.8	2.5	2.8	0.8	−5.6	−35.0
11	Petroleum & coal prods	−50.0	5.8	8.1	−5.7	1.8	−60.9	1.0
12	Chemical industry	−155.2	3.9	5.2	−26.0	23.7	6.1	−168.1
13	Ceramic & clay	−299.1	−7.2	9.7	−6.2	15.4	20.4	−331.1
14	Primary metal	−325.7	7.6	73.5	−102.1	25.5	15.7	−345.9
15	Metal products	−10.3	−0.0	1.2	−1.4	0.5	−0.7	−9.9
16	General machinery	−40.3	0.3	4.5	−0.4	3.0	−6.6	−41.2
17	Transport machinery	−11.3	1.2	4.2	−0.3	0.1	−2.4	−14.0
18	Electric machinery	−11.4	0.1	0.3	−0.3	0.6	−0.0	−12.1
19	Electronic & comm eqp	−9.2	0.1	0.0	−0.4	0.3	0.1	−9.4
20	Precision equipment	−1.5	−0.0	−0.1	−0.9	0.6	−0.3	−0.9
21	Other manufacturing	−2.5	−1.4	−0.7	−1.9	−1.7	−3.9	7.1
22	Power & heat supply	−605.9	75.5	120.4	−207.0	63.4	−991.1	332.9
23	Gas production & supply	−8.9	0.4	0.1	−0.0	0.0	−0.0	−9.3
24	Water supply	−0.3	0.1	0.0	−0.0	0.0	−0.3	−0.0
25	Construction	−14.8	−0.6	0.1	−0.1	0.1	1.2	−15.5
26	Trans & Comm	−124.5	13.7	67.3	−25.1	2.7	18.9	−201.8
27	Commerce & restaurant	−27.3	−4.4	3.1	0.1	1.1	4.8	−32.0
28	Other services	−42.9	4.3	4.8	−4.6	3.0	22.3	−72.7
	Total	−1982.7	91.3	314.2	−423.5	116.8	−1035.0	−1046.5

Source: Authors' calculation

Table 4.5 DPG analysis of CO_2 emissions by Industry (2012–2017) (Unit: Millions ton)

	Industry	DPG	Cons	Inv	Export	Import Sub	Tech Change	Emi Coeff
1	Agric, forestry & fisheries	−16.7	0.0	−8.0	−2.0	−0.4	−9.2	2.8
2	Coal	−104.2	1.1	1.5	−2.1	1.2	−15.7	−90.2
3	Crude oil & natural gas	−19.5	1.3	0.8	−1.2	2.8	−0.6	−22.7
4	Metal mining	−13.0	0.1	−0.0	−0.3	−0.7	−2.1	−9.9
5	Non−metallic mining	−19.7	0.1	1.5	−0.2	−0.1	−0.6	−20.3
6	Food & Tobacco	−67.6	6.2	−1.0	−1.1	−0.7	−1.8	−69.2
7	Fiber	−34.4	−0.5	−0.4	−2.3	−0.2	−3.2	−27.9
8	Clothing & leather	−10.1	−0.2	−0.0	−0.4	−0.1	0.1	−9.5
9	Wood processing & furniture	−12.9	0.1	0.2	−0.1	−0.1	−0.3	−12.8
10	Papermaking & printing	−38.7	1.1	−0.1	−1.7	0.3	−1.1	−37.3
11	Petroleum & coal prods	−57.9	5.0	4.4	−3.7	2.0	−32.5	−33.3
12	Chemical industry	−148.1	6.7	1.7	−10.4	−2.9	−12.0	−131.1
13	Ceramic & clay	−559.3	2.2	112.7	−21.9	3.4	−107.1	−548.6
14	Primary metal	−664.7	16.3	−5.1	−57.4	27.4	−660.2	14.4
15	Metal products	−15.2	0.1	−0.2	−0.4	0.1	0.2	−15.0
16	General machinery	−38.8	0.3	−5.9	−1.3	0.1	0.1	−32.1
17	Transport machinery	−21.2	0.7	−0.8	−0.5	0.1	0.4	−21.0
18	Electric machinery	−9.5	−0.0	−0.3	−0.2	0.0	0.0	−9.1
19	Electronic & comm eqp	−2.3	0.1	−0.0	−0.3	0.4	0.3	−2.7
20	Precision equipment	−1.1	0.0	−0.0	−0.0	0.1	0.1	−1.1
21	Other manufacturing	−3.2	0.1	−0.0	−0.0	0.5	−0.1	−3.5
22	Power & heat supply	−1189.8	43.2	60.1	−119.0	12.8	−642.4	−544.4
23	Gas production & supply	−1.6	0.0	0.0	−0.0	0.0	0.2	−1.8
24	Water supply	−0.5	−0.0	0.0	−0.0	0.0	−0.0	−0.5
25	Construction	−6.4	0.0	6.2	−0.1	−0.0	−0.7	−11.9
26	Trans & Comm	−74.8	46.0	57.4	−7.9	−0.4	93.7	−263.7
27	Commerce & restaurant	−27.6	0.7	0.6	−3.4	−0.5	9.9	−34.9
28	Other services	−47.6	8.5	5.0	−2.5	0.1	7.5	−66.3
	Total	−3206.6	139.2	230.4	−240.4	45.2	−1377.0	−2003.9

Source: Authors' calculation

and clay, and the chemical industry. Again, technological change and a decline in emission coefficients were the factors behind this, indicating that the decarbonization of the entire intermediate input structure has progressed further.

6 Concluding Remarks

This paper analyzed the factors behind the change in China's industrial structure and how such change relates to changes in CO_2 emissions.

We compiled fixed–price time series input–output tables for 2007–2012 and 2012–2017 based on the input–output tables published by the National Bureau of Statistics and applied the DPG analysis, which is a method that assumes a proportional growth economy in which each industry grows at the same rate, and decomposes the difference between the actual and proportional growth into various demand factors. The results obtained are summarized as follows.

First, consumption and investment were positive, and technological change and exports were negative factors in both periods. Later (2012–2017), the consumption factor expanded rapidly and became overwhelmingly positive. Conversely, although the investment factor was positive, the magnitude decreased. The Chinese economy was well known as an export–led economy; however, it changed from export–led to domestic demand–led, with a recent shift from investment–led to consumption–led.

Second, the leading growth industries in the first period (2007–2012) were transportation, chemical, electrical, and machinery industries, such as electronic and communication equipment. The factors were consumption, investment, and import substitution. In contrast, in non – manufacturing industries, transportation, communications, and other services expanded; however, growth industries shifted from manufacturing to service industries in the second period (2012–2017). Notably, technological change significantly affected the service industry's expansion, and the economy's service orientation is also progressing in China.

Finally, the DPG of CO_2 emissions was almost negative during both periods. The DPG of CO_2 was negative in energy-intensive industries such as primary metals, ceramic and clay, chemicals, petroleum and coal products, and non-energy-intensive manufacturing and service industries. The main drivers of the change were (energy-saving) technology change and emission factor change. Regarding the changes by industry, the electric power and heat supply industry recorded a significant negative figure, indicating that energy conservation (power saving) has progressed in the Chinese economy. Comparing the first (2007–2012) and second periods (2012–2017), the negative range of CO_2 emission DPG in the second expanded in each industry, confirming the acceleration of energy-saving technology changes.

Acknowledgements This work was supported by JSPS KAKENHI Grant Numbers JP19K12459.

References

Chen X, Guo J (2000) Chinese economic structure and SDA model. J Syst Sci Syst Eng 19(2):142–148

Chenery HB (1960) Patterns of industrial growth. Am Econ Rev 50(4):624–654

Chenery HB, Shishido S, Watanabe T (1962) The pattern of Japanese growth, 1914−1954. Econometrica 30:98–39

Fujikawa K (1996) Industrial structural change and its factors: experiences in Japan, Korea and Taiwan. Bus Econ 31:88–116. Institute of small business research and business administration, Osaka University of Ecoboimics. (in Japanese)

Fujikawa K (1997) An analysis of China's economic growth and energy/environmental issues, econometric approach to environmental problems, Ch1, Part1. Econ Anal 154. Economic and Social Research Institute. (in Japanese). https://www.esri.cao.go.jp/jp/esri/archive/bun/bun154/bun154a.pdf

Fujikawa K, Ninomiya S (1997) Changes in China's industrial structure and their factors. J Osaka Univ Econ 47(6):45–90. (in Japanese)

Guan D, Peters G, Weber C, Hubacek K (2009) Journey to world top emitter: an analysis of the driving forces of China's recent CO_2 emissions surge. Geophys Res Lett 36(4). https://doi.org/10.1029/2008GL036540

Hasebe Y (1995) Changes of Chinese economic structure and environmental load. Economia 46 (3):52–64. Yokohama National University. (in Japanese). https://ynu.repo.nii.ac.jp/index.php?action=pages_view_main&active_action=repository_action_common_download&item_id=2074&item_no=1&attribute_id=20&file_no=1&page_id=59&block_id=74

Japan External Trade Organization (JETRO) (2009) Economic stimulus measures by the Chinese government and their effects, business opportunities and risks for Japanese companies, Overseas Research Department, JETRO. (In Japanese). https://www.jetro.go.jp/ext_images/jfile/report/07000102/urgent-report_china090813.pdf

Jiang T (2008) The SDA analysis of industrial development in transitional China. Rev Ind Econ 7 (3):115–132. Shandong University. (in Chinese). https://kns.cnki.net/kcms/detail/detail.aspx?dbcode=CCJD&dbname=CCJDLAST1&filename=CYJP200803010&uniplatform=NZKPT&v=o7RtqQlFE6arxbhqIro7Sad692BEk4WLdKLV6qUXse-NY5YBGnpsedKw5JJJVcJ3

Jin J, Hasebe Y (2006) The factor analysis of the economic structure change in China: an analysis based on constant price I-O tables. Economia 57(2):19–28. Yokohama National University. (in Japanese). https://ynu.repo.nii.ac.jp/?action=repository_uri&item_id=2117&file_id=20&file_no=1

Li B (2015) Influences of industrial structural change on growth of Chinese economy from a viewpoint of inter−industrial relationships. Reg Econ Stud 26:29–40. Hiroshima University. (in Japanese). https://cres.hiroshima−u.ac.jp/kiyou/26/26_03.pdf

Liao M, Xu L (2017) IO-SDA model of CO_2 emissions and its empirical research. Stat Res 34 (7):62–70. (in Chinese). https://tjyj.stats.gov.cn/CN/10.19343/j.cnki.11-1302/c.2017.07.006

Ministry of Economy, Trade and Industry (2018) White paper on international economy and trade 2018. (in Japanese). https://www.meti.go.jp/english/report/data/wp2018/wp2018.html

Peng S, Zhang W, Sun C (2015) China's production-based and consumption-based carbon emissions and their determinants. Econ Res J 50(01):168–182. (in Chinese). http://www.erj.cn/en/IssueInfo.aspx?m=20150821085059340239

Peters G, Weber C, Guan D, Hubacek K (2007) China's growing CO_2 emissions: a race between increasing consumption and efficiency gains. Environ Sci Technol 41(17):5939–5944. https://doi.org/10.1021/es070108f
Teng J (1997) Factor analysis of China's economic growth and industrial structural change. Asia Keizai 38(2):44–61. (in Japanese)
World Meteorological Organization (2022) Global annual to decadal climate update. https://hadleyserver.metoffice.gov.uk/wmolc/WMO_GADCU_2022−2026.pdf
Ye Z, Jin J, Fujikawa K (2016) Changes in China's industrial structure and their factors. In: Kiyoshi F (ed) Input output analysis and applied general equilibrium analysis on the Chinese economy, Ch1. Horitsu Bunka Sha, pp 3–18. (in Japanese)
Zhang Y (2010) Supply-side structural effect on carbon emissions in China. Energy Econ 32 (1):186–193. https://doi.org/10.1016/j.eneco.2009.09.016

Statistical Data

CO_2 Emissions

Carbon Emission Accounts & Datasets. https://www.ceads.net/data/nation/

Price by Industry

National Bureau of Statistics, China statistical yearbook 2008-2019, China Statistics Press. https://spc.jst.go.jp/statistics/stats_index.html

Input Output Table

National Bureau of Statistics (ed) (2009) China input output table 2007. China Statistics Press
National Bureau of Statistics (ed) (2015) China input output table 2012. China Statistics Press
National Bureau of Statistics (ed) (2019) China input output table 2017. China Statistics Press

Chapter 5
Productivity and Eenergy Efficiency of Chinese Industries

Takatoshi Watanabe and Kiyoshi Fujikawa

Contents

Keywords China · CO_2 emissions · Energy efficiency · Input–Output Table · Total Factor Productivity (TFP)

Abbreviation

TFP Total Factor Productivity

T. Watanabe (✉) · K. Fujikawa
Faculty of Economics, Aichi Gakuin University, Nagoya, Aichi, Japan
e-mail: twata@dpc.agu.ac.jp; fujikawa@dpc.agu.ac.jp

K. Fujikawa (ed.), *Empirical Research on Environmental Policies in China*,
https://doi.org/10.1007/978-981-99-5957-0_5

1 Introduction

China's full-fledged economic growth began with the adoption of the Reform and Open Door Policy in 1978. This policy, which was promoted by then Vice President Deng Xiaoping for the introduction of capital and technology from abroad, was a modification of Mao Zedong's self-reliance policy. However, this policy of reform and liberalization has not always been well implemented. In the 1980s, immediately after the introduction of the Reform and Open Door Policy, there was an investment boom in China from developed countries, and the Chinese economy grew rapidly. However, this movement was suppressed by the Chinese government, which does not tolerate political freedom, resulting in the tragedy of the Tiananmen Square incident. The suppression of political freedom also reflected a sense that communism would be undermined by the economic power of the West among Chinese conservatives. Developed countries condemned the incident and imposed economic sanctions. As a result, foreign investment in China plummeted.

Figure 5.1 shows the economic growth rates of Japan and China. In 1989, there was a sharp decline in China's economic growth rate due to conservative modifications to the Reform and Open Door Policy. However, the southern provinces, where economic growth was beginning to take off, were dissatisfied with these backward steps. In response to these developments, Deng Xiaoping toured the south of China in 1992 and reemphasized the concept of a "socialist market economy" that guarantees freedom of economic activity (the so-called "Southern Tour Discourse"). The pro-reform and open-door movement once again took control of China's economic management as a result of this. Foreign investment then began to surge, and China once again recorded double-digit economic growth.

Undoubtedly, the Reform and Open Door Policy was the catalyst for China's "takeoff," and while there was a debate among China researchers in the 1990s about

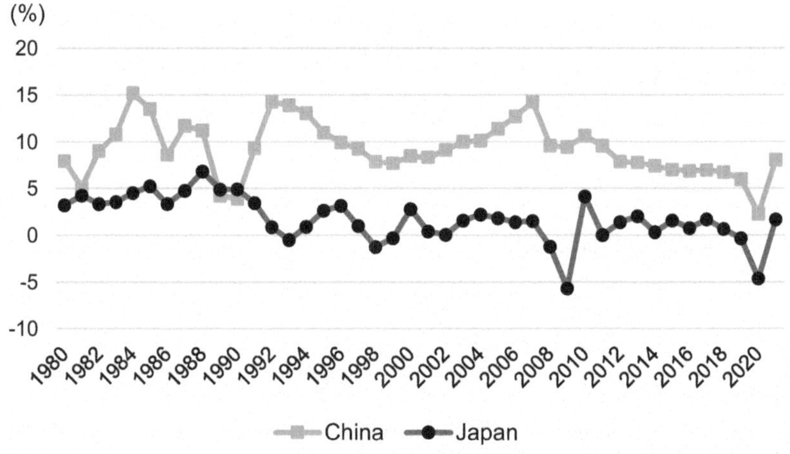

Fig. 5.1 GDP growth in Japan and China. Source: IMF—World Economic Outlook Databases

the "quality" of this economic growth, the World Bank (1993, 1997) evaluated the policy as a success and made optimistic forecasts that China's growth would continue. In contrast, Krugman (1994) argued that the high economic growth rates in the early years of economic development in socialist countries were due to the increased mobilization of factors of production and not increased productivity. Furthermore, he pessimistically predicted that China's production expansion would not be long-lasting.

However, if we look back at Japan's past experience, Japan's industrial structure became more sophisticated through the introduction of technology from advanced Western countries, and during the high-growth period, improvements in productivity emerged as the primary source of economic growth Saito (2000). In addition, it is widely recognized that the end of the high-growth period was partly attributable to the narrowing of the technology gap between Japan and the Western industrialized countries in the early 1970s. Given that China's economy has also achieved economic growth by leveraging foreign capital and technology, it is natural to expect that China will follow the same process as Japan in the years to come.

While there are numerous different assessments of the growth of the Chinese economy during the 1980s and 1990s, Fujikawa and Watanabe (2005) used the periods 1987–1992 and 1992–1997 to estimate the rates of productivity growth in different industries in China. Krugman's (1994) assertion was, in part, unfounded; while the increases in total factor productivity (TFP) in the 1980s were not observed to be significant, as will be discussed below, a large TFP growth area was observed in the 1990s, particularly in the industrial sectors. Much as was the case in Fujikawa and Watanabe (2005), one of the objectives of this study is to estimate productivity growth by the industrial sector in China in the 2000s. In other words, we will examine whether industries with higher productivity growth rates were also improving their energy efficiency and reducing their CO_2 emissions.

In the following sections of this study, Sect. 2 describes the model used in this paper to estimate productivity growth rates, and Sect. 3 introduces previous research on production in China. Section 4 presents the estimated results for China's productivity by industry, followed by a discussion of the relationship between the TFP growth rate and eco-efficiency in Sect. 5. Section 6 provides an interpretation of these estimates, and Sect. 7 presents our conclusions.

2 Model

Following Solow (1957) and Denison (1967), it is widely recognized that the growth of productivity is one of the most important factors for economic growth. There are various measurement methods for productivity growth depending on the form of the aggregation function of the inputs. The trans-log function, i.e., the quadratic function of the logarithm, is now frequently used. A trans-log function is defined in Eq. (5.1), where the amount of output is denoted by Y and the amount of the ith input is denoted by X_i, $(i = 1, \cdots, n)$.

$$ln\,Y = a_0 + \sum a_{1i}\,ln\,X_i + \sum\sum a_{2ij}\,ln\,X_i \cdot ln\,X_j \quad (5.1)$$

As the trans-log production function is a quadratic function, the growth of output from period 0 to period 1 can be defined as follows, using the quadratic lemma:

$$ln\,Y_1 - ln\,Y_0 = \frac{1}{2}\sum_i \left(\frac{\partial\, ln\,Y_1}{\partial\, ln\,X_{1i}} + \frac{\partial\, ln\,Y_0}{\partial\, ln\,X_{0i}}\right)(ln\,X_{1i} - ln\,X_{0i}) \quad (5.2)$$

Or,

$$ln\,Y_1 - ln\,Y_0 = \frac{1}{2}\sum_i \left(\frac{X_{1i}}{Y_1}\frac{\partial Y_1}{\partial X_{1i}} + \frac{X_{0i}}{Y_0}\frac{\partial Y_0}{\partial X_{0i}}\right)(ln\,X_{1i} - ln\,X_{0i}) \quad (5.2')$$

China's economic system is not necessarily based on the principle of profit maximization under conditions of perfect competition. However, if we could apply the marginal theory that assumes each input's real wage to be equal to its marginal productivity, Eq. (5.2′) can be transformed as follows:

$$ln\,Y_1 - ln\,Y_0 = \frac{1}{2}\sum_i \left(\frac{X_{1i}}{Y_1}\frac{q_{1i}}{p_1} + \frac{X_{0i}}{Y_0}\frac{q_{0i}}{p_0}\right)(ln\,X_{1i} - ln\,X_{0i}) \quad (5.3)$$

Or,

$$ln\,Y_1 - ln\,Y_0 = \frac{1}{2}\sum_i (w_{1i} + w_{0i})(ln\,X_{1i} - ln\,X_{0i}) \quad (5.3')$$

In the above-described equations, the symbols p and q_i are the prices of the output and the ith input, respectively, and w_i is the nominal input share of the ith input.

This equation states that, as long as the form of the production function is unchanged, the growth rate of output is the same as the weighted average of the growth rate of each input, where the weights are determined by each input's nominal share. However, the equality in Eq. (5.3′) does not necessarily hold when the observed data are substituted in because the shape of the production function usually varies with changes of time. Therefore, we define the difference between the left-hand side and the right-hand side of the equation as caused by efficiency change. That is to say, when the left-hand side is larger, the cause is an efficiency improvement. Eq. (5.4) is the definition of the growth of TFP.

$$TFP = (ln\,Y_1 - ln\,Y_0) - \frac{1}{2}\sum_i (w_{1i} + w_{0i})(ln\,X_{1i} - ln\,X_{0i}) \quad (5.4)$$

3 Previous Research on TFP in China

3.1 Previous Research Covering the Period Up to the 1990s

Young's (1994) work estimated the TFP growth rates for four newly industrializing countries (NIEs), namely, Hong Kong, Singapore, South Korea, and Taiwan, and compared them with the results of Elias's (1990) (for South American countries) and Christensen et al. (1980) (for developed countries). Similar to Krugman (1994), this study argued that the economic growth of newly industrializing Asian countries was generally a "factor of production mobilization type" and "these TFP growth rates are neither higher than those of South American countries nor greater than the past experience of advanced economies."" Young (2003) subsequently estimated China's TFP growth rate using self-adjusted data. The author stated that the TFP growth rate of the industrial sectors was about half (1.4%) of that estimated using official data (3.0%) when adjusted data were used and stated that while a TFP growth rate of 1.4% is certainly large, it is not surprisingly so (Table 5.1).

Furthermore, Table 5.2 shows the results of the World Bank (1997) estimates. As shown in the right column of Table 5.2, the share of the effects of input growth and

Table 5.1 Comparison of TFP growth rates in Asia, developed countries, and Latin America

Developing countries	Period	Annual rate (%)	Advanced countries	Period	Annual rate (%)
Hong Kong	1966–91	2.3	Canada	1947–73	1.8
Singapore	1966–90	- 0.3	France	1950–73	3.0
South Korea	1966–90	1.6	West Germany	1950–70	3.7
Taiwan	1966–90	1.9	Italy	1952–73	3.4
Argentina	1940–80	1.0	Japan	1952–73	4.1
Brazil	1950–80	2.0	Netherlands	1951–73	2.5
Chile	1940–80	1.2	Britain	1955–73	1.9
Colombia	1940–80	0.9	The USA	1947–73	1.4
Mexico	1940–80	1.7			

Source: Young (1994)

Table 5.2 World Bank comparison of TFP growth rates for China, Japan, the U.S., and South Korea

		Average annual growth rate (%)				Factor share (%)	
Country	Period	GDP	Capital equipment	Human capital	Labor force	Inputs	Technological change
China	1978–1995	9.4	8.8	1.6	2.4	71	29
USA	1950–1992	3.2	3.2	1.1	1.6	65	35
Japan	1960–1993	5.5	8.7	0.3	1.0	70	30
South Korea	1960–1993	8.6	12.5	3.5	2.4	79	21

Source: World Bank (1997)

Table 5.3 TFP growth in China by industrial sector

1987–1992		Output Growth	Contribution of inputs: Int. inputs	Labor	Capital	TFP growth
10	Metals	7.84	10.24	0.84	0.76	−5.61
11	General Machinery	12.15	10.22	1.04	0.41	1.44
12	Trans. Equipment	20.28	17.64	0.91	0.73	3.37
13	Elec. Appliances	11.37	9.30	0.95	0.61	1.43
14	Other Manufacturing	22.51	20.04	1.11	1.16	2.87
15	Construction	5.97	3.04	2.24	0.25	0.88
16	Elec. Gas & Water	15.39	11.92	0.26	6.04	−1.32
17	Transportation	6.06	9.83	0.71	2.77	−10.92
18	Trade & Catering	22.88	19.80	3.94	1.21	1.15
19	Services	11.91	11.36	2.29	2.69	−4.82
Average		10.35	9.07	1.13	1.00	−0.34
1992–1997		Output Growth	Contribution of inputs			TFP growth
10	Metals	10.42	10.36	0.06	0.33	−0.31
11	General Machinery	9.35	4.75	−0.44	0.21	5.61
12	Trans. Equipment	22.10	18.91	0.59	0.65	4.96
13	Elec. Appliances	22.74	19.64	0.21	0.87	5.06
14	Other Manufacturing	19.99	9.55	0.60	0.42	13.19
15	Construction	10.54	12.96	0.94	0.13	−5.43
16	Elec. Gas & Water	4.28	10.38	−0.03	1.52	−12.33
17	Transportation	14.81	5.21	1.08	2.45	8.75
18	Trade & Catering	5.22	3.68	1.98	0.84	−1.11
19	Services	11.31	8.25	2.35	1.36	0.58
Average		11.75	9.46	0.27	0.72	2.31

Source: Fujikawa and Watanabe (2005)

technological progress among the factors contributing to China's economic growth from 1978 to 1995 was 7:3. This is comparable to the TFP growth rates of Japan, the U.S., and South Korea. This result implies that China's TFP growth rate after reform and liberalization was 2.73% per year, which is a relatively high figure.

Fujikawa and Watanabe (2005) estimated productivity growth rates by industrial sector for China using data for the period 1987–1997. Their estimation results are shown in Table 5.3. The growth rate of output in each industry was reasonably high in both periods; however, the TFP growth rate underlying this was clearly different between the two periods. In the period 1987–1992, little TFP growth was observed, and Krugman's (1994) assertion may have been true for China at that time. In the period 1992–1997, however, improvements in TFP expanded in the machinery industry, including in transportation machinery and electrical equipment, as

Table 5.4 The Lehman Brothers collapse: China's TFP growth rate

	Value-added growth	Capital input (%)	Labor input (%)	TFP(%)
1981–1991	8.81	5.82	1.12	1.86
1991–2001	8.85	7.00	1.12	0.72
2001–2007	11.37	9.45	0.59	1.32
2007–2012	9.22	10.39	0.25	−1.42

Source: Wu and Liang (2017)

technologies from abroad were actively introduced, and it can no longer be argued that the economy grew solely by mobilizing total factors of production.[1]

3.2 *Previous Research Covering the 2000s and Subsequent Years*

Wu and Liang (2017) found a substantial TFP growth rate of 1.86% in the 1980s, as shown in Table 5.4, which they mainly attributed to agricultural reforms (mainly the introduction of market forces in rural areas). While the TFP growth rate in the 1990s was sluggish at 0.72% due to a lack of progress in the reform of state-owned enterprises, the TFP growth rate for the period up to the collapse of Lehman Brothers was also estimated to be high. This was attributed to the increase in exports following the country's accession to the WTO in 2001. However, after the collapse of Lehman Brothers, the TFP growth rate turned negative from 2007 to 2012.

Brandt et al. (2020) estimated TFP in China before and after the collapse of Lehman Brothers in 2008 and found that the TFP growth rate was 3.5% from 1979 to 1988, immediately after the adoption of the Reform and Open Door Policy, and 2.8% from 1999 to 2008, prior to the Lehman Brothers collapse, which was remarkably high. However, the growth rate significantly declined to 0.7% in the period 2009–2018 following the crisis. It was pointed out that although the Chinese economy continued to grow even after the global financial crisis, the main reason for this was an increase in capital from public investment and the limited contribution of TFP.

Wei et al. (2017) estimated the TFP for China from 1979 to 2015 and found that the main factor contributing to the growth of the Chinese economy was an increase in capital, with only a limited contribution from a quantitative increase in labor. Moreover, the TFP growth was positive until 2008 (before the Lehman Brothers

[1] Fujikawa and Watanabe (2005) examined the relationship between TFP growth in the Chinese economy and the introduction of foreign capital and export activities of foreign-invested firms and confirmed that the larger the share of foreign-invested firms, the higher the TFP growth rate of the industry in question, and the higher the TFP growth rate of an industry, the higher its export share in total sales. It was verified that, in the 1980s and 1990s, foreign-invested enterprises in China played a role in raising the productivity of export industries and boosted export-led economic growth.

crisis) but turned negative thereafter. The aggressive fiscal policy after the collapse of Lehman Brothers stimulated capital accumulation but did not lead to structural reforms to increase efficiency, which is similar to the assessment of Brandt et al. (2020).

What these studies have in common is that they find that the TFP growth rate declined after the collapse of Lehman Brothers and the ensuing crisis. As discussed below, these results are consistent with the results presented in this paper. The reasons for this will be discussed below.

4 Estimated TFP by Industrial Sector in China

4.1 Data Used

This study uses three connected input–output tables for China for the years 2007, 2012, and 2017. To make the production and input volumes comparable over time, we created a Linked Input-Output Tables (real table) from these three input–output tables, valued at 2007 prices.[2] Price coefficients from the China Statistical Yearbook were used for the price data. For the industrial sector classification, the production sectors (columns) were merged into 28 sectors, and the intermediate input sectors (rows) were merged into two sectors: energy and non-energy. Production factors are defined as labor and capital. The labor input was calculated using the China Population & Employment Statistics Yearbook, and the capital input was calculated using the capital depletion allowance in the input–output table. However, since the capital depreciation allowance is a nominal value, it was converted to real 2007 prices using the fixed capital investment price index in the China Statistical Yearbook.[3]

4.2 Estimated TFP Growth by Industrial Sector in China

The estimation period for the TFP growth rate in this paper covers 2007–2017 and is split into two: the first (2007–2012) and second halves (2012–2017) of the period. In the first half of 2008, China's economy was in the midst of the global recession triggered by the collapse of Lehman Brothers, and export-led economic growth was showing signs of faltering. Therefore, in the fall of 2008, the Chinese government announced a 4 trillion RMB stimulus package (public investment) to compensate for the decline in foreign demand. In the second half of the period, the Chinese economy

[2]For details, please refer to Chap. 4 of Ye and Fujikawa (2023).

[3]This is not the ideal method but due to data limitations, this method was used as the best available solution.

Table 5.5 TFP growth in China by industrial sector (2007–2012) (Unit:%)

	Industry	Output	Material inputs	Energy inputs	Labor inputs	Capital inputs	TFP
1	Agric, forestry & fisheries	3.97	2.92	0.10	-2.09	0.16	2.88
2	Coal	7.18	4.18	2.90	0.25	0.57	-0.71
3	Crude oil & natural gas	-1.36	1.73	-0.84	-1.78	0.65	-1.13
4	Metal mining	8.69	6.25	0.57	-0.11	0.76	1.22
5	Non-metallic mining	-0.08	-0.22	0.25	-0.51	-0.00	0.41
6	Food & tobacco	9.62	7.55	0.07	0.24	0.18	1.60
7	Fiber	2.03	1.51	-0.02	-0.75	0.02	1.28
8	Clothing & leather	8.89	7.38	0.07	0.04	0.25	1.15
9	Wood processing & furniture	7.52	6.16	0.10	-0.29	0.44	1.11
10	Papermaking & printing	10.01	7.96	0.31	0.32	0.40	1.02
11	Petroleum & coal prods	4.22	1.62	4.08	-0.11	0.03	-1.40
12	Chemical industry	9.64	7.84	0.49	0.33	0.32	0.66
13	Ceramic & clay	9.38	7.59	0.95	1.88	0.60	-1.64
14	Primary metal	9.18	7.14	0.55	-0.32	0.46	1.35
15	Metal products	8.71	7.66	0.51	0.63	0.27	-0.35
16	General machinery	9.27	8.49	0.06	0.19	0.30	0.22
17	Transport machinery	10.64	9.05	0.06	-0.46	0.25	1.75
18	Electric machinery	10.05	8.64	0.14	0.26	0.23	0.78
19	Electronic & comm eqp	9.45	7.46	0.03	0.57	0.13	1.26
20	Precision equipment	2.47	1.37	-0.01	-0.55	-0.00	1.66
21	Other manufacturing	-17.33	-17.61	-0.28	-2.37	-0.31	3.24
22	Power & heat supply	5.73	1.06	3.14	-0.05	-0.06	1.65
23	Gas production & supply	12.62	4.67	6.31	0.09	0.42	1.13
24	Water supply	3.78	2.70	0.04	-0.19	0.51	0.72
25	Construction	9.27	8.75	0.19	1.48	0.16	-1.30
26	Trans & Comm	10.98	8.48	0.29	0.23	0.62	1.35
27	Commerce & restaurant	9.85	4.79	-0.01	1.43	0.76	2.88
28	Other services	11.86	7.58	0.07	0.73	1.71	1.77
	Total	9.01	6.72	0.47	0.06	0.50	1.25

Source: Authors' estimation

began to transition from having a foreign-demand-led structure to one that is driven by domestic demand (consumption)-led economic growth. As represented by the terms "new normal" and "moderately prosperous society," the Chinese economy at this time was facing the challenge of shifting from a period of high economic growth to one of stable growth and beyond.

TFP Growth Rate from 2007 to 2012

Table 5.5 shows the TFP growth rates (%) by industry for the first half of the period. The shaded cells in the table for output and TFP represent the top five industry sectors. The average increase in output was as high as 9.0% across all industries. The industrial sectors with particularly high rates of increase were gas production and supply, other services, and transportation and communication, all of which are services sectors. In the industrial sector, transport machinery and electric machinery were the top exporters, while papermaking and printing, chemical industry, and electronic and communication equipment also showed high growth rates.

The rate of increase in TFP was estimated to be generally positive, averaging 1.3% across all industries. The high TFP growth rates were estimated in other manufacturing and transpor machinery in the manufacturing industry, and in commerce and restaurant, and other services in the service industry. The rate of increase in TFP is defined as the difference between the rate of change in output and the rate of change in the factors of production, and labor input was found to be the main contributor to the rate of increase in TFP. Underlying the increase in TFP in many industries was a decrease in labor input due to improvements in labor efficiency.

Estimated Results from 2012 to 2017

Table 5.6 shows the rate of increase in TFP by industry for the second half of the period. The output growth rate for all industries sharply declined to 5.8% from 9.0% in the first half. The industries with the highest rates of increase in output were

Table 5.6 TFP growth in China by industrial sector (2012–2017) (Unit:%)

	Industry	Output	Material inputs	Energy inputs	Labor inputs	Capital inputs	TFP
1	Agric, forestry & fisheries	2.84	1.30	0.03	-2.55	-0.02	4.08
2	Coal	2.63	-0.81	0.39	-3.53	-0.14	6.73
3	Crude oil & natural gas	7.11	-2.53	-0.69	-1.36	3.77	7.92
4	Metal mining	0.83	-1.83	-0.48	-2.31	-0.15	5.60
5	Non-metallic mining	6.43	4.07	0.89	-3.08	0.85	3.71
6	Food & tobacco	6.16	5.16	0.11	-0.29	0.09	1.09
7	Fiber	0.48	0.33	0.01	-0.70	-0.07	0.92
8	Clothing & leather	3.81	4.14	0.06	-0.62	0.01	0.22
9	Wood processing & furniture	5.18	4.83	0.08	0.01	0.02	0.24
10	Papermaking & printing	4.98	4.40	0.21	-0.10	0.09	0.38
11	Petroleum & coal prods	2.99	0.46	3.06	-0.18	-0.10	-0.25
12	Chemical industry	4.75	3.09	0.38	-0.13	0.16	1.24
13	Ceramic & clay	5.64	3.86	0.87	-0.50	0.23	1.18
14	Primary metal	-0.10	-1.19	0.18	-0.54	-0.03	1.47
15	Metal products	5.47	3.77	0.28	-0.18	0.18	1.42
16	General machinery	2.00	1.36	0.04	-0.25	0.01	0.85
17	Transport machinery	5.55	4.45	0.06	-0.10	0.02	1.11
18	Electric machinery	4.07	3.14	0.05	0.19	0.14	0.54
19	Electronic & comm eqp	7.53	6.90	0.09	0.31	0.21	0.01
20	Precision equipment	7.71	6.43	0.19	-0.08	0.03	1.14
21	Other manufacturing	7.56	4.26	0.03	-1.03	0.88	3.42
22	Power & heat supply	3.36	1.21	0.65	-0.17	0.93	0.75
23	Gas production & supply	9.96	1.69	9.68	0.16	0.86	-2.43
24	Water supply	5.00	3.23	0.69	-0.79	0.40	1.47
25	Construction	7.70	7.09	0.15	0.33	0.03	0.10
26	Trans & Comm	10.22	5.34	0.31	1.54	3.34	-0.30
27	Commerce & restaurant	7.21	5.06	0.17	1.17	1.70	-0.90
28	Other services	7.60	5.54	0.13	1.48	0.07	0.39
	Total	5.78	4.04	0.28	0.05	0.42	1.00

Source: Authors' estimation

transportation and communications, gas production and supply, other service industries, and construction. These are the same sectors that also showed high output growth rates in the first half of the period. This indicates that China's economy has become more service-oriented since the year 2000. In the industrial sector, the output growth rate for transportation machinery and electric machinery was large in the first half of the period; however, the growth rate of these industries slowed down in the second half of the period. Transportation and electric machinery are export industries that have previously supported the growth of the Chinese economy but their exports have declined, which is thought to be a result of the global recession.

As in the first half, the TFP growth rates in the second half of the period were generally positive, although certain industrial sectors recorded negative rates of growth. In the second half of the period, TFP rose at a higher rate in the mining industry, including coal, crude oil and natural gas, and metal mining. The main exporting industries were transportation machinery, electric machinery, and electronic and communication equipment, which decreased from 1.75% to 1.11%, 0.78% to 0.54%, and 1.26% to 0.01%, respectively. In the second half of the period, the major contribution to the increase in TFP was continued improvements in the efficiency of labor input, and the number of industries with negative labor-input effects increased over the first half of the period.

5 The TFP Growth Rate and CO_2 Emissions of the Chinese Economy

The total amount of output expanded by 1.7 and 1.4 from 2007 to 2012 and 2012 to 2017, respectively. However, CO_2 emissions were 1.4 and 1.0 times higher in the first and second halves of the period, respectively. Generally, the expansion of industrial output is thought to increase energy use and CO_2 emissions. These figures indicate a decrease in CO_2 emissions per unit of production. Accordingly, this section examines the relationship between the TFP growth rate, energy productivity, and CO_2 emissions per unit of energy.

Table 5.7 shows the TFP growth rates (restated), energy productivity change rates, and per-unit CO_2 emissions contribution rates by industry. The shaded cells in industrial sectors indicate the top five energy-input coefficients and energy-intensive sectors; the shaded cells in TFP-increase rate and energy productivity change rate indicate the top five industrial sectors. Furthermore, the shaded cells in the CO_2 emissions per-unit contribution rate indicate the top five industrial sectors in terms of reduction rate. The shaded cells for TFP-increase rate and energy productivity change rate indicate the top five industrial sectors.

Table 5.7 TFP growth, energy productivity change, and CO2 emissions contribution per unit by industrial sector

	Unit : %	TFP growth rate (Annual average)		Energy productivity (Annual average)		CO2 emissions contribution per unit	
		2007-12	2012-17	2007-12	2012-17	2007-12	2012-17
1	Agric, forestry & fisheries	2.88	4.08	-5.86	-0.21	-0.10	0.02
2	Coal	-0.71	6.73	-5.18	1.25	0.04	-0.13
3	Crude oil & natural gas	-1.13	7.92	5.66	19.31	0.01	-0.06
4	Metal mining	1.22	5.60	6.93	4.74	-0.02	-0.07
5	Non−metallic mining	0.41	3.71	-2.34	-0.91	0.02	-0.04
6	Food & tobacco	1.60	1.09	7.46	-1.77	-0.53	-0.70
7	Fiber	1.28	0.92	2.97	0.10	-0.17	-0.24
8	Clothing & leather	1.15	0.22	2.76	-2.82	-0.29	-0.40
9	Wood processing & furniture	1.11	0.24	5.53	2.20	-0.10	-0.27
10	Papermaking & printing	1.02	0.38	2.97	-1.25	-0.21	-0.29
11	Petroleum & coal prods	-1.40	-0.25	-1.56	-1.40	0.00	-0.06
12	Chemical industry	0.66	1.24	6.18	0.25	-0.69	-0.58
13	Ceramic & clay	-1.64	1.18	4.07	-0.98	-0.13	-0.21
14	Primary metal	1.35	1.47	3.99	-2.36	-0.27	0.01
15	Metal products	-0.35	1.42	1.37	0.74	-0.17	-0.35
16	General machinery	0.22	0.85	9.14	0.29	-0.58	-0.56
17	Transport machinery	1.75	1.11	8.80	0.37	-0.41	-0.79
18	Electric machinery	0.78	0.54	2.07	0.37	-0.54	-0.69
19	Electronic & comm eqp	1.26	0.01	9.51	-0.89	-1.05	-0.60
20	Precision equipment	1.66	1.14	3.19	-5.31	-0.04	-0.09
21	Other manufacturing	3.24	3.42	-5.20	8.23	0.13	-0.07
22	Power & heat supply	1.65	0.75	0.35	2.39	0.05	-0.06
23	Gas production & supply	1.13	-2.43	0.97	-4.77	-0.05	-0.05
24	Water supply	0.72	1.47	3.97	1.16	-0.00	-0.01
25	Construction	-1.30	0.10	-0.88	-1.48	-0.51	-0.41
26	Trans & Comm	1.35	-0.30	1.29	-1.49	-0.37	-0.54
27	Commerce & restaurant	2.88	-0.90	12.97	-2.55	-0.38	-0.43
28	Other services	1.77	0.39	11.94	-3.74	-1.28	-1.22
	Total	1.25	1.00	4.53	1.92	–4.00	–5.97

Source: Authors' estimation

5.1 *Energy Productivity Change Rate*

Energy productivity is "domestic production per unit of energy input." The higher the number is, the higher the energy efficiency is. The energy efficiency of all industries increased by 4.5% in the first half of the period, while improvements in efficiency declined in the second half of the period, limiting growth to 1.9%.

5.2 Contribution Ratio of Per-Unit CO_2 Emissions

The contribution rate of per-unit CO_2 emissions is defined as the product of the rate of change in CO_2 emissions per unit of domestic production value versus the share of the relevant industry sector in domestic production value. The per-unit CO_2 emissions of all industries decreased by 4.0% and 6.0% in the first and second halves of the period, respectively, indicating that the impact of CO_2 emission reductions is expanding.

5.3 TFP and Energy Productivity Change Rate

The TFP growth and energy productivity were positive in many industrial sectors in both the first and second periods, and the relationship between the two is shown in the scatter plots. Figures 5.2 and 5.3 show the relationship between the rate of increase in TFP and the rate of change in energy productivity for the first and second halves of the period, respectively.[4]

The regression lines in Figs. 5.2 and 5.3 are both right ascending, indicating a positive correlation between the rate of increase in TFP and the rate of change in energy productivity. However, the correlation coefficient were 0.54 and 0.43 in the

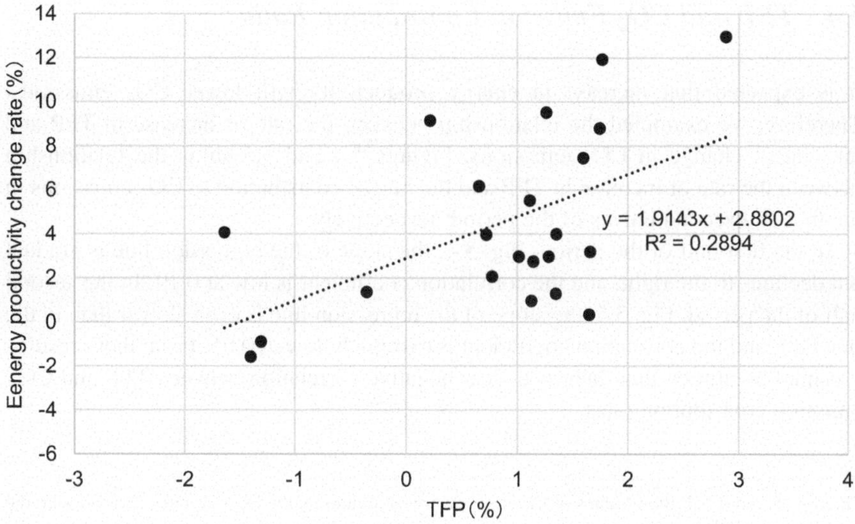

Fig. 5.2 TFP and energy productivity change (2007–2012). Source: Authors' estimation

[4]Other manufacturing and agriculture, forestry and fisheries were excluded.

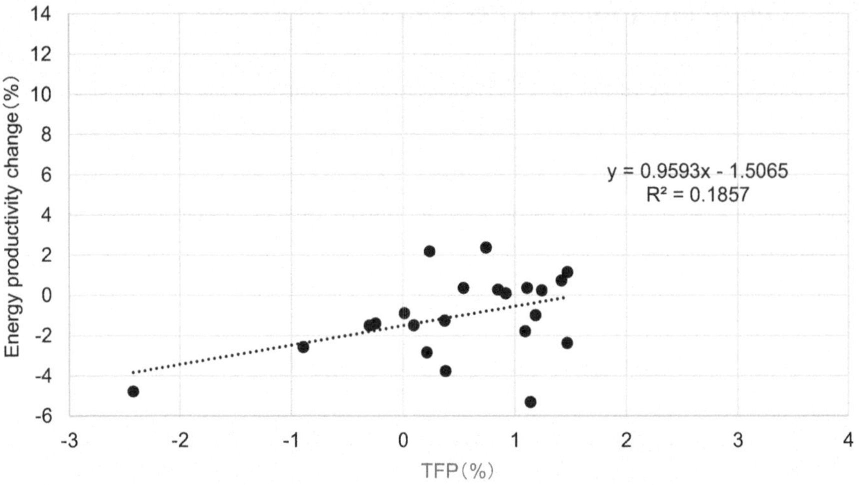

Fig. 5.3 TFP and energy productivity change (2012–2017). Source: Authors' estimation

first and second halves, respectively, indicating that the correlation in the second half was not as strong as in the first half.[5]

5.4 *TFP and CO_2 Emission Contribution Ratio*

It is expected that increase in energy productivity will lower CO_2 emissions. Therefore, we examined the relationship between the rate of increase in TFP and the rate of change in CO_2 emissions. Figures 5.4 and 5.5 show the relationship between the rate of increase in TFP and the rate of contribution to CO_2 emissions in the first and second halves of the period, respectively.

In the first half of the period, Fig. 5.4, the slope of the regression line is gradual but declines to the right, and the correlation coefficient is low at 0.19. In the second half of the period, Fig. 5.5, the slope of the regression line is even slower than in the first half, and the correlation coefficient is extremely low at 0.05. From these results, it cannot be argued that there is a clear negative correlation between TFP and CO2 emission contribution rates.[6]

[5]In Figs. 5.2 and 5.3, the t-values of the slopes were estimated by the least-squares method with the explained variable as the rate of change in energy productivity and the explanatory variable as the rate of increase in TFP and the values were 2.85 in the first period and 2.13 in the second period. Both were significant at the 5% significance level.

[6]In Figs. 5.4 and 5.5, the t-values of the slopes were − 0.87 and − 0.22 for the first and second periods, respectively, when the explained variable was the contribution rate of CO_2 emissions and the explanatory variable was the rate of increase in TFP, estimated by the least-squares method. Both were not significant at the 5% significance level. This means that the relationship between the

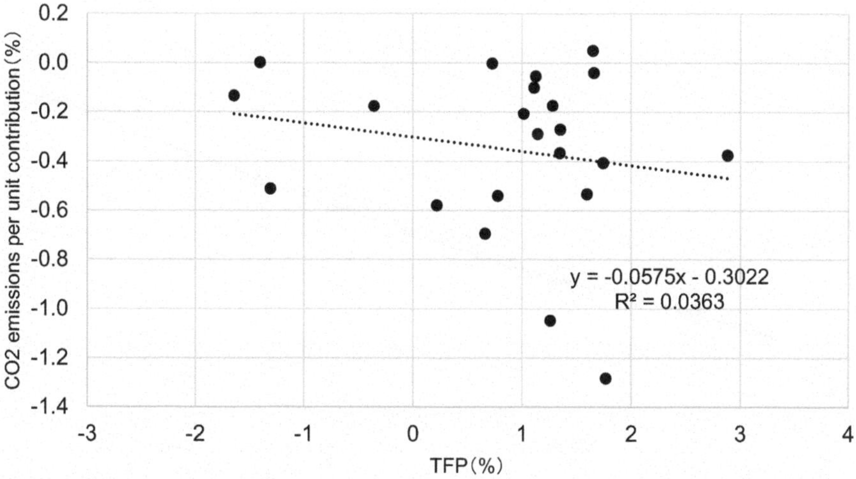

Fig. 5.4 TFP and per-unit CO2 emissions contribution (2007–2012). Source: Authors' estimation

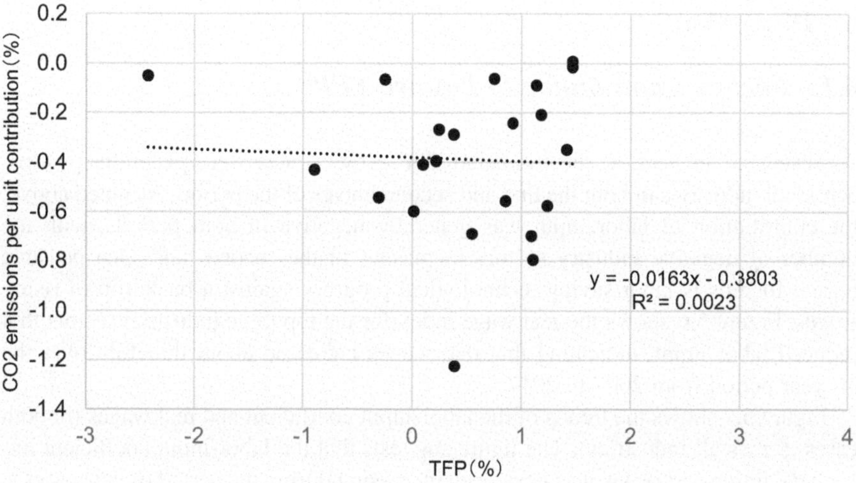

Fig. 5.5 TFP and CO2 emissions contribution per unit (2012–2017). Source: Authors' estimation

rate of increase in TFP and the rate of contribution to CO_2 emissions cannot be argued to have a clear negative correlation.

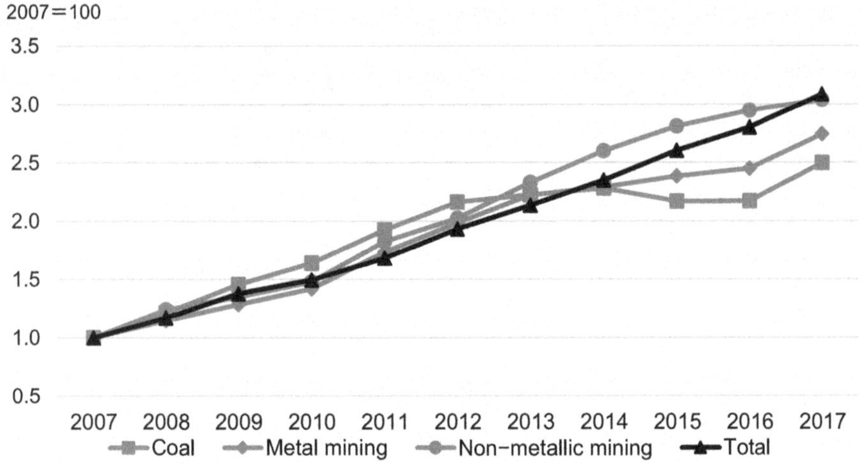

Fig. 5.6 Real wage rates in China (2007–2017). Source: Created by the authors based on China labor statistical yearbook and China statistical yearbook (each year)

6 Discussion

6.1 Factors Contributing to Positive TFP

As discussed in Sect. 4, the estimated TFP growth rates were generally positive across all industries in both the first and second halves of the period. As noted above, the contribution of labor input was generally negative in both periods, with the number of negative industry sectors increasing in the second half. One possible reason for this is labor-saving technological progress against a backdrop of rising wages. Figure 5.6 shows the real wage index for the top three industrial sectors that reduced labor input, indicating that real wages increased about threefold over the 10-year period from 2007 to 2017.

Figure 5.7 shows the trends of the labor-input coefficient and real wages (in both cases, across all industries). The figure suggests that the labor-input coefficient has been declining and real wages have been increasing during the period of the analysis. Keeping in mind that the labor-input coefficient is the inverse of labor productivity, Fig. 5.7 shows that labor productivity and real wages rose simultaneously in the Chinese economy during this period. It can be argued that firms may have responded to a rise in real wages with labor-saving technological advances during this period.

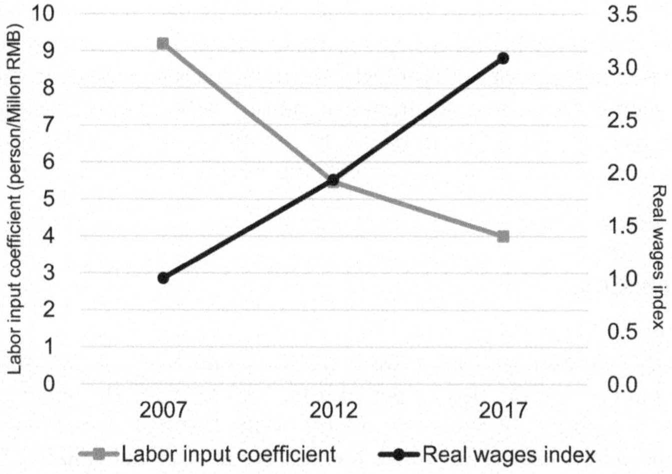

Fig. 5.7 Labor-input coefficient and real wage rate in China. Source: Created by the authors based on China labor statistical yearbook and China statistical yearbook (each year)

6.2 *Factors Contributing to the Decline in the Rate of Increase in TFP*

In many industrial sectors, TFP-increase rates in the second half of the period were lower than those in the preceding one. The reasons for this are discussed below. The first is a decline in production; the Chinese economy, which is highly dependent on exports, saw its output decline due to the global recession triggered by the Lehman Brothers collapse at the start of the second period. This decline in exports can be presumed to have reduced the rate of increase in production output.

The second reason is likely to be an overestimation of capital and labor input. Capital investment immediately made after the Lehman Brothers collapse resulted in an excess of capital, and the capacity utilization rate is thought to have declined. However, since the capacity utilization rate was not considered in the calculations in this paper, capital input may have been overestimated. This, in turn, may have led to the observed decline in the rate of increase in TFP. The same is true for labor input, which may also be overestimated.

6.3 *Relationship Between TFP-Increase Rate and CO_2 Emissions Per Unit*

Despite improvements in energy efficiency, the relationship between the rate of increase in TFP and CO_2 emissions per unit of production was not clearly negatively correlated. We believe that this is due to the fact that only direct emissions were used as a measure of CO_2 emissions in this study. For example, if energy productivity increases in an industry but the increase is due to a reduction in electricity input and not a reduction in the input of fossil fuels, electricity consumption does not directly emit CO_2. Hence, CO_2 emissions from that industry will not be reduced using this calculation method. The CO_2 emissions data should have taken into account the indirect CO_2 emissions that are also emitted during electricity production.

7 Conclusion

In this study, we estimated the growth rate of productivity by industry for the Chinese economy over the 10-year period from 2007 to 2017. The output increased at a relatively high rate of about 9% and 6% in the first and second halves of the period, respectively, while TFP growth was positive in many individual industries in both the first and second halves. It was also generally positive across all industries over the full period. The introduction of labor-saving technologies is thought to have contributed to this increase in productivity in both halves of the period. In the latter half of the period, exports from China also declined due to the global recession following the collapse of Lehman Brothers. As a result, the production value of major export industries declined, and the rate of increase in TFP also fell from 1.75% to 1.11%, 0.78% to 0.54%, and 1.26% to 0.01% for transportation machinery, electric machinery, and electronic and communication equipment, respectively.

To clarify the relationship between productivity growth and eco-efficiency, we examined the relationship between TFP growth rate, energy productivity, and CO_2 emissions per unit of production. The rate of increase in TFP and energy productivity confirmed that, to a certain extent, productivity gains increased energy productivity. However, we could not find a clear correlation between the rate of increase in TFP and per-unit CO_2 emissions, which may be due to the fact that direct CO_2 emissions cannot be used in isolation to calculate the total amount of CO_2 emissions.

Finally, this section discusses issues for future research. First, the accuracy of the TFP estimation needs to be improved. Since the TFP growth rate in this study is defined as the difference between the rate of change in output and the rate of change in each factor of production input, the accuracy of the TFP growth rate estimation depends on the accuracy of the estimation of factors of production inputs. The capital input used in this study is based on the accounting allowance for capital depletion, which may not accurately capture the actual capital services input. The labor-input data should also be captured on a person-hour basis, taking into account the number of hours worked.

The relationship between the TFP growth rate and CO_2 emissions also needs to be estimated more precisely, and the indirect CO_2 emissions from electricity production should also be considered instead of only using direct CO_2 emissions.

Acknowledgments This work was supported by JSPS KAKENHI Grant Numbers JP19K12459 and JP21H04941.

References

Brandt L, Litwack J, Mileva E, Wang L, Zhang Y, Zhao L (2020) China's Productivity Slowdown and Future Growth Potential. In: Policy Research Working Paper No.9298. Word Bank

Christensen L, Cummings D, Jorgenson D (1980) Economic Growth, 1947-1973: An International Comparison. In: Kendrick JW, Vaccara B (eds) New Developments in Productivity Measurement, NBER Studies in Income and Wealth, vol 41. University of Chicago Press, Chicago, pp 595–698

Elias V (1990) Sources of Growth: A survey of Seven Latin American Economies. Institute of Contemporary Studies Press

Fujikawa K, Watanabe T (2005) Productivity growth in the Chinese economy by industry. China & World Economy, Institute of World Economics and Politics, Chinese Academy of Social Sciences, pp 56–67

Krugman P (1994) The Myth of the Asia's Miracle. Foreign Aff 73(6):62–78

Saito M (2000) The Japanese Economy. World Scientific Publishing

Solow R (1957) Technical Change and the Aggregate Production Function. Rev Econ Stat 39:312–320

Wei S, Xie Z, Zhang X (2017) From "Made in China" to "Innovated in China": Necessity, Prospect, and Challenges. J Econ Perspect 31(1):49–70

World Bank (1993) The East Asian Miracle: Economic Growth and Public Policy. Oxford University Press, Oxford

World Bank (1997) China 2020. World Bank

Wu H, Liang D (2017) Accounting for the Role of Information and Communication Technology in China's Productivity Growth, RIETI Discussion Paper Series, 17-E-111. The Research Institute of Economy, , Trade and Industry

Young A (1994) The Tyranny of Numbers: Confronting the Statistical Realities of the East Asian Growing Experience, NBER Working Paper Series No.4680, National Bureau of Economic Research. Republished in Quarterly Journal of Economics, August 1995:614–680

Young A (2003) Gold into Base Metals: Productivity Growth in the People's Republic of China during the Reform Period. J Polit Econ 111:1220–1261

Part III
Empirical Research on Environmental Policies in China

Chapter 6
Optimal Location for Large-Scale Wind Farms in China

Jiayang Wang

Contents

Keywords Renewable energy · Wind power · Optimal location · Power curtailment · Ultra-high voltage

Abbreviation

NDRC National Development and Reform Commission
NEA National Energy Administration
SGCC State Grid Corporation of China
SPC State Power Corporation
UHV ultra-high voltage

J. Wang (✉)
Faculty of Economics, Aichi Gakuin University, Nagoya, Aichi, Japan
e-mail: jywang@dpc.agu.ac.jp

K. Fujikawa (ed.), *Empirical Research on Environmental Policies in China*,
https://doi.org/10.1007/978-981-99-5957-0_6

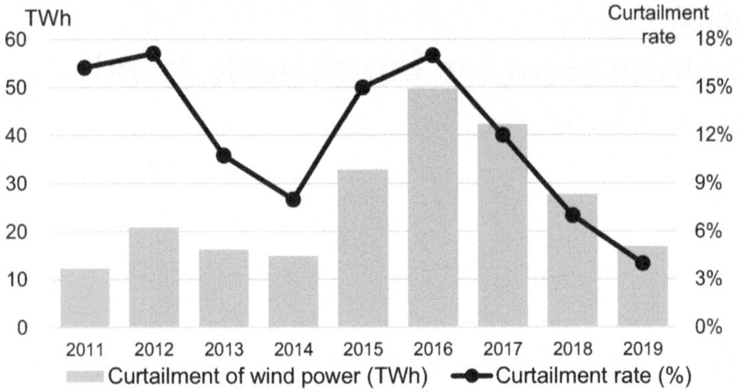

Fig. 6.1 Power curtailment amount and rate of China's wind power. Source: Created by the author based on data from the NEA (each year)

1 Introduction

In 2006, the Chinese government enacted the "Renewable Energy Law" and began to introduce renewable energy power generation in earnest. Special efforts have been made to promote wind power generation, and between 2006 and 2010, the annual growth rate of installed wind power capacity exceeded 100%. As a result, in 2011, China's installed wind power capacity base surpassed that of the United States, to become the largest in the world.

While, on one hand, wind power generation installations are spreading rapidly, a curious phenomenon has occurred: wind power output is being suppressed. The reasons for this are as follows: Most wind farms in China have been built in the northern and western regions of the country where land prices are low and there is an abundance of wind resources. However, these regions are far from the eastern coastal regions, where electricity demand is high, and a strong power grid is needed to utilize the power in the eastern coastal regions, and so the development of the power grid has not kept up with the rapid growth of wind power plants.

Figure 6.1 shows the amount of wind power curtailment since 2011 and its ratio to the generated capacity (power curtailment ratio), which was quite high at 16.2% and 17.1% for wind power in 2011 and 2012, respectively. It then dropped temporarily due to the efforts of the grid operators to adjust supply and demand, but because of the limits to their abilities to do so, it reached 17% again in 2016. Since then, the power curtailment rate has been on a downward trend.

On a regional basis, wind power development was concentrated in the northwestern and northern regions, and this made it difficult to adjust the supply-demand balance within the region. As shown in Table 6.1, the high rate of wind power output curtailment in Gansu, the Xinjiang Uyghur Autonomous Region, Jilin Province, Heilongjiang, and Inner Mongolia is a result of this.

Table 6.1 Power curtailment rate of China's wind power by regional (%)

Regional	2013	2014	2015	2016	2017	2018
Gansu	31.0	11.0	39.0	43.0	33.0	19.0
Xinjiang	4.3	15.0	32.0	38.0	29.0	22.9
Jilin	15.7	15.0	32.0	30.0	21.0	6.8
Heilongjiang	11.5	12.0	21.0	19.0	14.0	4.4
Inner Mongolia	15.2	9.0	18.0	21.0	15.0	10.3

Note: Data for 2014–2017 are only available in integer values
Source: Created by the author based on data from the National Energy Administration (NEA) (n.d., various years)

A renewable energy policy assessment conducted by the European Commission (Ragwitz et al. 2007) pointed out that immature promotion schemes are potentially a major risk affecting the spread of renewable energy generation and concluded that more emphasis should be placed on the stability and reliability of the system as a whole. In fact, the introduction of feed-in tariffs in Europe has been criticized repeatedly by the Federation of Electric Power Companies of Europe and others for concentrating wind power generation projects in remote areas with low economic activity, land prices, and population density, and imposing unnecessary cost on consumers (Kainou (2008)).

The total cost of supplying electricity from large-scale renewable energy power plants can be divided into two categories: power generation cost and transmission cost (Kainou (2009)). The power generation cost include the construction and operation cost of power plants, and the major difference between regions is the cost of expropriating land for the construction of power plants. The cost of land acquisition declines as the location of the power plant moves further away from the city center. On the other hand, the further away from the city center the power plant is located, the longer the transmission distance becomes and thus the higher the transmission cost becomes. Therefore, there is an optimal distance from the city center at which the total cost is minimized. However, if the power generation and transmission businesses are different, the power producer only considers the cost of power generation and does not need to consider the cost of transmission. In other words, power producers will build power plants far from the city center where land prices are low, regardless of the total cost of supplying electricity in the area in which it is consumed.

In China, the power generation and transmission businesses were separated in 2002, and a feed-in tariff for wind power was introduced in 2009 (NDRC(2009)). This requires transmission companies to purchase wind-generated electricity, with the transmission companies bearing the cost of transmission and able to recover the cost out of the profits gained from the sale of electricity. While power producers are building large numbers of power plants in remote areas where wind conditions are favorable, China's electricity prices are regulated by the government, so changes in transmission cost are not fully reflected in electricity prices. Therefore, the transmission operators could not operate their businesses if they purchased power

Table 6.2 Regional grid of China

Regional grid	Provinces
Northeast	Heilongjiang, Liaoning, Jilin, Inner Mongolia(East)
North China	Beijing, Tianjing, Shandong, Hebei, Shanxi
Northwest	Shaanxi, Ningxia, Gansu, Qinghai, Xinjiang, Tibet
Inner Mongolia	Inner Mongolia (West)
East China	Shanghai, Jiangsu, Zhejiang, Anhui, Fujian
Central China	Chongqing, Jiangxi, Henan, Hubei, Hunan, Sichuan
South China	Guangdong, Guangxi, Hainan, Guizhou, Yunnan

Source: Created by the author based on State Grid Corporation of China (SGCC) HP (2023)

generated by renewable energy. In this way, the fact that wind farms were built in remote areas due to differences in cost-sharing entities is one of the reasons for the curtailment of wind power output.

In the past, China's power grid was operated at the regional grid level as shown in Table 6.2. Power supply and demand was regulated within the regional grid network. Therefore, there was little inter-regional power grid flexibility, and the high-voltage transmission lines were mainly short, at less than 300 km in length. However, as mentioned above, wind power plants were concentrated in the western and northern regions, where wind resources are abundant, making it difficult to coordinate the supply and demand for power within the region. As a result, wind power generation in the western and northern regions had to be suppressed.

The most effective way to solve the power curtailment problem is to transmit wind-generated power from the western and northern regions to the eastern coastal region, where demand for electricity is high. However, since the transmission distance in this case exceeds 500 km, the cost of conventional transmission methods will be high, and the transmission business will only be profitable if the power is transmitted at higher voltages. Therefore, the Chinese government has implemented a plan to build ultra-high voltage (UHV), long-distance transmission lines to transmit wind-generated power from the western and northern regions to the eastern region.

Based on the above, this study considered how the optimal location of wind power generation could be changed by using transmission lines with different voltages. We conclude that the construction of a transmission network at higher voltage would considerably lower the urban supply price of wind-generated electricity in the western and northern regions. In other words, the profitability of wind-generated electricity in the western and northern regions has improved.

2 China's Electric Power Supply System

In 2002, the Chinese government reformed its power supply system by separating the generation and transmission of power. This separation of power generation and transmission is one of the reasons for the suppression of wind power output in the northern part of the western region, as already mentioned.

As shown in Table 6.2, China's power grid is roughly divided into six regions: Northeast, North China, Northwest, Inner Mongolia, East China, Central China, and South China, and the supply and demand of electricity was adjusted between these regions. As mentioned above, the concentration of wind power development in the northwest and north regions has made it difficult to adjust the supply-demand balance within the regions.

In order to solve the problem of power curtailment in these regions, the Chinese government has plans to transmit wind power from the Inner Mongolia Grid, the Northwest Grid, and the Northeast Grid to the eastern coastal regions where demand for electricity is high. However, because the transmission distance to the eastern region exceeds 500 km, UHV transmission lines are required. The UHV transmission system can compress the current and control power loss even when transmitting power over long distances. Conversely, transmitting power over long distances exceeding 500 km using normal high-voltage 220 kV transmission lines would result in large power losses and very high transmission cost.

This study compared the use of two transmission lines, one 220 kV and the other 750 kV, to examine how the optimal location for transmitting wind-generated power varies between existing high-voltage transmission lines and new UHV transmission lines.

3 A Model for Minimizing the Location Cost for Power Plants

3.1 Kainou Model

This study uses a model for minimizing the location cost for renewable energy power plants (Kainou (2009)). Figure 6.2 illustrates the Kainou model.

Wind power generation requires an overwhelmingly large land area compared to thermal power plants of the same capacity. This is due to technical restrictions such as the need for large spacing between wind turbines, in addition to the extremely low energy density of wind power. Therefore, in addition to the wind conditions, the land value of the site itself is critically important in determining the location of wind power generation.

In this study, it is assumed that the Beijing metropolitan area (Jing-jin region) is a demand center for electric power. The metropolitan area, along with the Yangtze

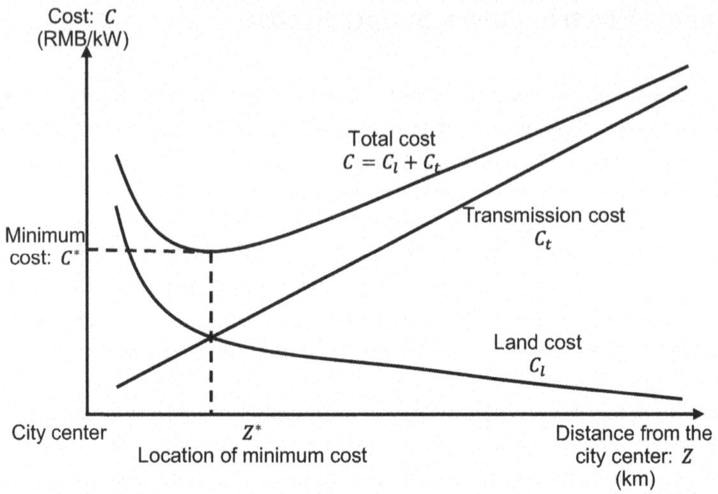

Fig. 6.2 Illustration of Kainou model. Source: Created by the author based on Kainou (2009)

River Delta and the Pearl River Delta, is the most densely populated and industrially developed region in China.

The total cost of a power plant consists of the cost of expropriating the land on which the plant is to be built (land cost), the cost of the generator itself and its installation, and the cost of the transmission facilities (transmission cost). However, the generator and installation cost are excluded from the cost considered in the power plant location model in this section because they are considered to be independent of the location of the power plant.

In general, as the distance (Z) from the city center (the location of electricity demand) increases, the land cost (C_l) decreases because competition for land use decreases and land prices become cheaper. The incline of the decline is initially steep, but this is expected to gradually slow down as the distance increases. Next, transmission cost (C_t) are expected to increase proportionally with the increase in the distance of the transmission line extension. Therefore, the transmission cost (C_t) are assumed to be a straight line with a constant term. Therefore, the total cost of constructing a power plant ($C = C_t + C_l$) is U-shaped as shown in Fig. 6.2, and this exists at a distance (Z^*) from the city center that achieves the minimum cost (C^*).

3.2 *Estimation of Land Cost*

Land cost is expressed as the product of land area and land price, and land price is considered to depend on the distance from the city center, as described above. Although the data is somewhat old, for land prices, we used 210 items of industrial land data around the metropolitan area from the China Land Market Network (2010).

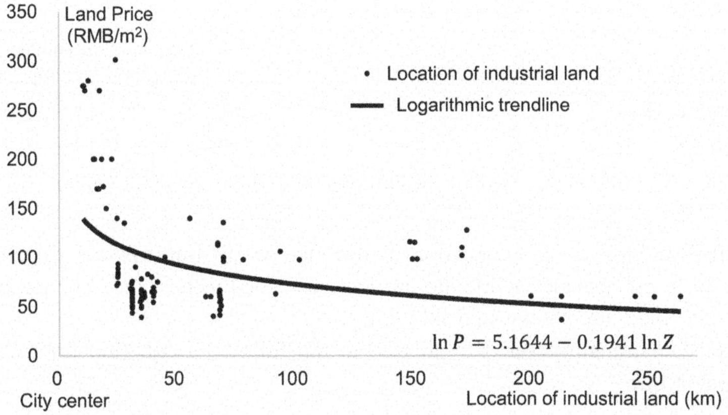

Fig. 6.3 The industrial land price and distance from the city center. Source: Created by the authors based on the data from the LandChina.com (2011)

However, this data does include industrial parks. Land prices in industrial parks are higher than those in surrounding areas because of the infrastructure for roads and electricity. This makes them unsuitable for power plant construction. Therefore, in this study, the industrial parks were excluded from the sample, and the remaining 154 data items were used to estimate the relationship between land price and distance (shown in Fig. 6.3). The estimation results are shown in Eq. (6.1), where P is the land price (RMB per square meter) and Z is the distance of the land from the city center.

$$\ln P = \underset{(26.23)}{5.1644} - \underset{(-3.91)}{0.1941}\ \ln Z \tag{6.1}$$

The coefficient of distance (ln Z) was significant at a 95% degree of significance. The adjusted R-square with degrees of freedom was 0.801. According to this estimation, land price decreases by about 0.19% when the distance of land from the city center increases by 1%. The land cost can be expressed as follows.

$$\begin{aligned} C_l &= A \cdot P = 49.6 \cdot \exp(5.1644 - 0.1941 \cdot \ln Z) \\ &= 49.6 \cdot e^{5.1644} \cdot Z^{-0.1941} \end{aligned} \tag{6.2}$$

A is a constant value for the area of land required for the power plant which was set to 49.6 m^2/kW in this study with reference to Kainou (2009).

3.3 Estimation of Power Cost

Transmission cost (c_t) consist of transmission facility construction and maintenance cost (c_{tn}) and substation facility construction and maintenance cost (c_{ts}) The former

Table 6.3 The parameters used in the transmission cost

Cost	Transmission voltage	
	220 kV	750 kV
C_{tn} (RMB/kW/km)	5.7	1.1
C_{ts} (RMB /kW)	135.3	149.7

Source: Created by the authors based on the data from SPC(2001) and SGCC(2008)

is assumed to increase in proportion to the increase in transmission distance (Z: distance from city center), while the latter are assumed to be fixed. The parameters used in the model are shown in Table 6.3.

The cost of transmission is expressed in the following equation for a 220 kV transmission line.

$$C_t = 135.3 + 5.7\ Z \tag{6.3}$$

If the transmission line is 750 kV, this can be obtained as follows:

$$C_t = 149.7 + 1.1\ Z \tag{6.4}$$

Therefore, we can solve the following minimization problem to find the location (distance from the city center) where the cost of the power plant will be lowest.

$$\underset{0 \leq Z}{\mathrm{Min}} C(Z) = \underset{0 \leq Z}{\mathrm{Min}} (C_l(Z) + C_t(Z)) \tag{6.5}$$

4 Estimation Results

When the transmission line is 220 kV, the least expensive location for a large-scale wind farm is 118 km from the city center. Figure 6.4 shows the locations of wind farms and the optimal location areas as of 2010. As shown in the figure, most of the wind farms in this region are located farther away from the optimal location area. Under the Chinese system, wind farm operators do not bear transmission cost, so their only concerns are wind conditions and land prices. This has led to wind farms being built far apart from urban centers.

Next, we estimated the minimum cost location for a case involving a 750 kV transmission line: the cost of 750 kV transmission is about 20% of that of 220 kV transmission, so the incline of the straight line in Fig. 6.2 representing the cost of transmission is quite moderate. However, since land prices are taken from 2010, the land cost curve in Fig. 6.2 is invariant. Therefore, the minimum cost point for 750 kV transmission is farther away from the metropolitan area than for 220 kV transmission.

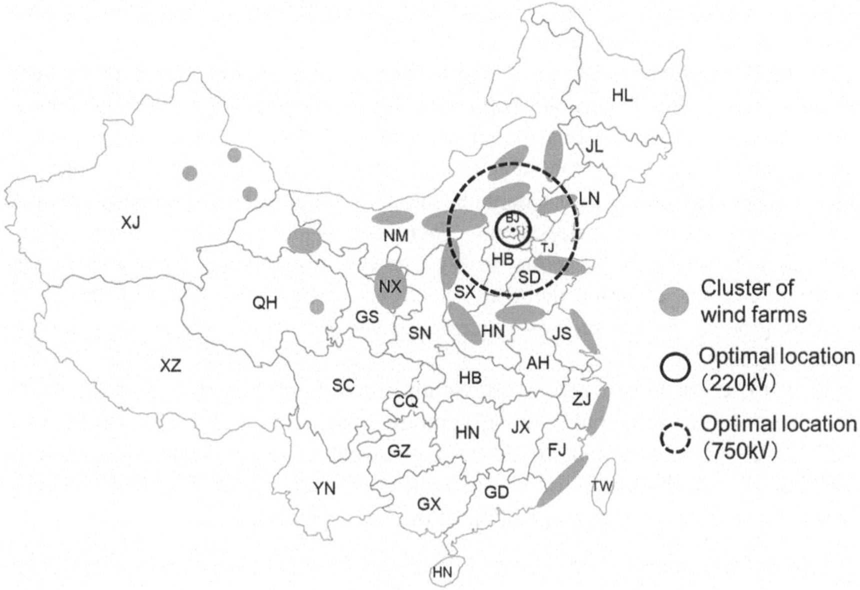

Fig. 6.4 Optimal location and current status of wind power in China. Source: The cluster of wind farms based on the data from people.cn(2021). The optimal location based on the author's simulation

The distance with the lowest total cost of power generation using 750 kV transmission lines is 471 km (Fig. 6.4). 750 kV transmission lines eliminate the cost bottleneck for wind power development in the Inner Mongolia grid, the Northeast grid, and the Northwest grid, and ensure the economic viability of transmission projects to the metropolitan area.

5 Situation and Issues Surrounding Long-Distance Transmission Lines

China is building UHV transmission lines for long-distance transmission at a rapid pace to solve the shortage of transmission capacity. According to the China Electricity Council,[1] the total distance of transmission lines above 750 kV in China in 2010 was about 10,657 km. However, by 2020, this will be 66,682 km, a six-fold increase. On the other hand, the total distance of transmission lines below 750 kV has only increased by a factor of 1.5, and they have been replaced by transmission

[1] China Electricity Council, "2011 Basic Electricity Statistical Data List" and "2020 Basic Electricity Statistical Data List" (China Electricity Council n.d., 2021).

lines with a relatively high voltage. Therefore, since 2016, there have been improvements in wind power suppression.

While UHV transmission lines tend to ensure the economic efficiency of long-distance transmission, there are some challenges in operating UHV transmission lines. First, the large transmission capacity of UHV transmission lines requires a large amount of electricity for stable transmission. On the other hand, the output of renewable energy power is not stable because the output of a single power plant is not large and is weather-dependent. Therefore, when operating a high-voltage transmission line, it is necessary to take power from several renewable energy power plants and combine them into a large capacity, and then add these to a regulating power source such as a thermal power or hydroelectric power source[2]to ensure stability.

In 2014, operation of the "Hami South-Zhengzhou" transmission line, which integrates wind and thermal power generation, began. The 14th Five-Year Plan for 2020–2025 includes a plan to develop renewable energy power generation in combination with hydroelectric and thermal power generation. This is a necessary measure to ensure the stable operation of the UHV power grid.

6 Conclusion

This study introduces a cost minimization model for the locating of large-scale wind power plants. Using this model, the optimal location of wind power plants in the metropolitan area of China (Jing-jin region) was examined by referencing, as an example, the change in the optimal location of wind power plants due to the increase in transmission voltage (which can be interpreted as a technological advancement in the electric power industry).

In China, where the power generation and transmission businesses are separate, the wind power promotion policy allows power producers to build power plants but not to bear the cost of transmission, so power plants are located in remote areas where the cost of land expropriation is low. However, for the utilities that transmit power to urban areas with high electricity demand, longer transmission lines increase cost. On the other hand, while transmission and distribution companies are obliged to purchase electricity generated from renewable energy sources, because the supply price of electricity is regulated, the increase in transmission cost cannot be directly reflected in the supply price. This has led to transmission and distribution companies refusing to allow renewable energy-generated electricity to be connected to the grid. If left unchecked, this could have put the brakes on the spread of renewable energy generation.

[2]Power sources necessary to maintain the supply-demand balance by adjusting the imbalance between supply and demand of electricity caused by load fluctuations and unexpected problems at power plants, and other factors.

Transmission of wind-generated power built in western and northern China to the eastern coastal areas, where demand for electricity is high, is an effective solution, but transmission over existing 220 kV transmission lines is too costly. This study examined the economic feasibility of transmitting wind-generated power at 750 kV, which is the highest voltage available. As a result, it was found that the point where the cost is minimized is far from the metropolitan area (expanded from about 120 km to about 500 km), and that transmission from Inner Mongolia and the northeastern region to the metropolitan area is also economically feasible.

As of 2020, the total distance of China's UHV transmission lines above 750 kV is about six times greater than in 2010. The development of the UHV grid is alleviating the problem of wind-generated electricity power curtailment in the northwest region. However, it should not be forgotten that the use of wind-generated power poses other problems. Wind power is an unstable power source as it is affected by weather and other factors. Therefore, it is necessary to develop a regulating power supply with thermal and hydroelectric power generation to transmit wind power at very high voltages. The Chinese government must build a power supply system capable of stably transmitting renewable energy power at very high voltage in order to further promote the spread of renewable energy.

Acknowledgements This work was supported by JSPS KAKENHI Grant Numbers JP19K12459372 and JP21H04941.

References

China Electricity Council (2021) The Annual Report on China's Electrification Development in 2020. China building materials Press. (in Chinese)

China Electricity Council (n.d.) Basic Electricity Statistical Data List. (various years, in Chinese)

Kainou K (2008) Current Situation and Problems in European Common Energy Policies, RIETI Policy Discussion Paper Series 08-J-001. pp 1–49. (in Japanese). https://www.rieti.go.jp/en/publications/summary/08040004.html

Kainou K (2009) Economic Considerations on Renewable Electricity Transmission/Distribution/Storage Cost Assistance System, RIETI Policy Discussion Paper Series 09-J-001, pp 1–43. (in Japanese). https://www.rieti.go.jp/jp/publications/summary/09010002.html

LandChina.com (2011) Announcement of land transfer of Beijing. (in Chinese)

National Development and Reform Commission (NDRC) (2009) Perfecting wind power feed-in tariffs policy. (in Chinese). https://chinaenergyportal.org/perfecting-wind-power-feed-in-tariffs-policy/

National Energy Administration (NEA) (n.d.) Wind power installations and production by province. (various years, in Chinese). https://chinaenergyportal.org/filter/?_sf_s=%E9%A3%8E%E7%94%B5%E5%B9%B6%E7%BD%91%E8%BF%90%E8%A1%8C%E6%83%85%E5%86%B5. Last Accessed 21 Jan 2023

People.cn (2021) Overview of the distribution of wind turbines in our country. (in Chinese). https://baijiahao.baidu.com/s?id=1706083211378947959&wfr=spider&for=pc. Last Accessed 31 Jan 2023

Ragwitz M, Held A, Resch G, Faber T, Haas R, Huber C, Coenraads R, Voogt M, Reece G, Morthorst EP, Jensen GS, Konstantinaviciute I, Heyder B (2007) OPTRES: Assessment and optimization of renewable support schemes in the European electricity market, final report.

European Commission. https://energy.ec.europa.eu/assessment-and-optimisation-renewable-energy-support-schemes-european-electricity-market-optres-0_en
State Power Corporation (SPC) (2001) Measures for the Calculation of Operation and Maintenance Cost of Power Grids. State Power Corporation. (in Chinese)
State Grid Corporation of China (SGCC) (2008) SGCC Labor quota standard. State Grid Corporation of China. (in Chinese)
State Grid Corporation of China Home Page. Organizational Structure. (in Chinese). http://www.sgcc.com.cn/html/sgcc_main_en/col2017112321/column_2017112321_1.shtml. Last Accessed 31 Jan 2023

Chapter 7
Initial Allocation of Emissions Trading Among Sub-regions in China

Yiyi Ju and Kiyoshi Fujikawa

Contents

Keywords Initial allocation · Emissions trading scheme · CO_2 · Gross regional product · Marginal abatement cost

Abbreviations

CEADs	Carbon Emission Accounts and Datasets
ETS	Emission Trading Scheme
GRP	Gross Regional Product
ICAP	International Carbon Action Partnership
MAC	Marginal abatement cost
UNICEF	United Nations International Children's Emergency Fund

Y. Ju (✉)
Waseda University, Shinjuku, Tokyo, Japan
e-mail: juyiyi@aoni.waseda.jp

K. Fujikawa
Aichi Gakuin University, Nagoya, Aichi, Japan
e-mail: fujikawa@dpc.agu.ac.jp

K. Fujikawa (ed.), *Empirical Research on Environmental Policies in China*,
https://doi.org/10.1007/978-981-99-5957-0_7

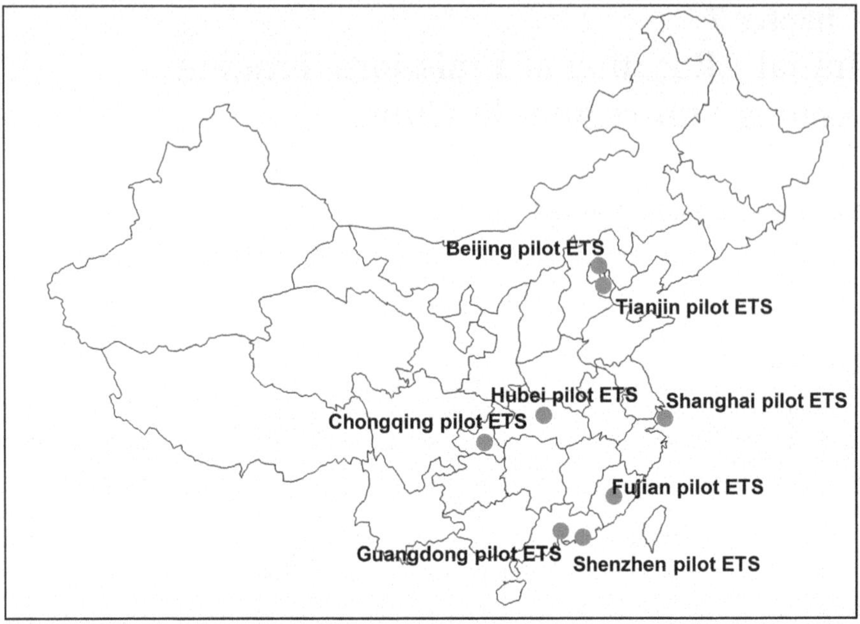

Fig. 7.1 ETS map of China. Source: Authors

1 Introduction

In 2021, China's national CO2 Emission Trading Scheme (ETS) started operating after three years of preparation period. The ETS currently regulates more than 2000 companies from the power sector, which emit 26,000 t-CO_2 per year. It is estimated to cover more than 4 billion t-CO_2, accounting for over 40% of national carbon emissions (ICAP 2021). Simultaneously, regional markets, originally constituting regional pilot markets, are operating parallel to each other in Beijing, Chongqing, Fujian, Guangdong, Hubei, Shanghai, Shenzhen, and Tianjin, as depicted in Fig. 7.1. Currently, the national ETS is intensity-based with an ex-post cap adjustment based on actual levels of production. During the period in which both regional and national ETSs were in operation concurrently, the original ETS pilot regions in China faced both demand-driven and policy-induced carbon leakages. To avoid such leakages, it was essential to adjust the initial allocation of allowances according to regional emission flows. In this study, we provided some recommendations to adjust the allocation of carbon emission allowances by adopting multiple effort-sharing approaches.

Numerous effort-sharing approaches based on various principles have been proposed. These principles include four basic tenets: responsibility, capability, equality, and cost-effectiveness. In this study, five effort-sharing approaches based on various principles were examined to provide a reference framework for the allocation of initial carbon emission allowances among different sub-regions in China.

2 Model and Data Sources

2.1 Model

Höhne et al. (2014) proposed the four distinct equity principles listed below.

1. Responsibility
   This principle considers the historical contribution of the region or country to global emissions or global warming.
2. Capability
   This principle is frequently referred to as the "right to development." Developing countries or regions could make less ambitious reduction efforts to secure their basic energy demands.
3. Human rights
   This principle recognizes our shared humanity and the equality of all human beings as constituting an equal claim to global collective goods (i.e., equal individual rights to atmospheric space).
4. Cost-effectiveness
   This principle considers the mitigation potential. In general, if there is considerable scope for reduction, marginal abatement costs (MACs) are low.

In this study, we considered and contrasted two perspectives on the theory of responsibility. The first aspect was the consumption of fossil fuels. Regions that consume fossil fuels are accountable for their direct contribution to CO_2 emissions. However, because fossil fuels are utilized to manufacture various end products, the regions in which the final products are consumed indirectly contribute to these CO_2 emissions. The former responsibility is referred to as "production responsibility," and the latter as "consumption responsibility." This study examined five distinct types of equity principles. Table 7.1 summarizes the relationships between the equity principles and their justifications.

The illustration of the emission reduction goal and initial emission allowances is shown in Fig. 7.2. We assume that each entity may originally receive its initial emission allowances that are equal to 90% of its annual emission in the previous year. Namely, the emission reduction goal of each entity would be 10% (the number was set at a higher value than reality to better show the difference that may be caused by adopting different allocation approaches). According to different effort-sharing principles, the responsibility of emission reduction would be adjusted, and the rest of the emission allowances would be the final initial allocation.

For data availability and simplicity of discussion, we assumed that the initial allocations of the 30 provinces in China were the actual CO_2 emissions in 2021. In

Table 7.1 Five emission allocation approaches

Approach	Effort-sharing principle	Justification
Energy consumption- based	Production Responsibility	Allocations of emission allowances based on the emissions generated from final energy consumption
Final consumption-based	Consumption Responsibility	Allocations of emission allowances based on the emissions resulting from the final consumption of goods and services
Equity	Right to development	Allocations of emission allowances based on population proportions
Ability to pay	Capability	Allocations of emission allowances depending on the capacity to bear the obligations
Cost-optimal	Cost-effectiveness	Allocations of emission allowances based on the least expensive options from marginal abatement cost (MAC) curves

Source: Authors, van den Berg et al. (2019), and Höhne et al. (2014)

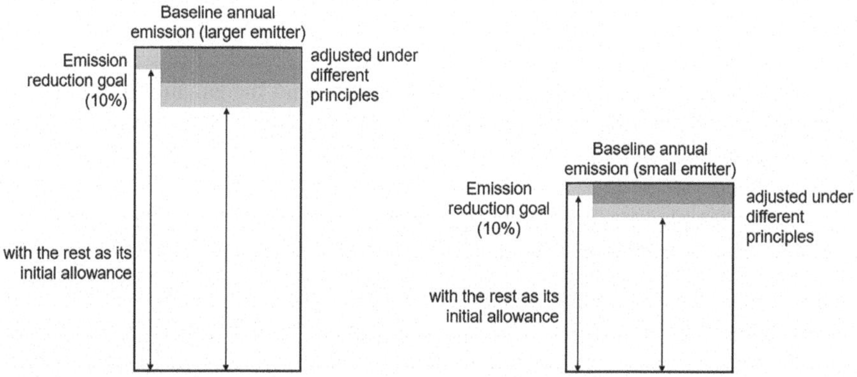

Fig. 7.2 Illustration of the emission reduction goal and initial emission allowances

addition, we calculated the necessary adjustment amount for the initial allocations for each approach.

The CO_2 emissions (that should be reduced in the target year) generated from energy use $\mathbf{e}_{\mathrm{pr}}$, is formulated as Eq. (7.1):

$$\mathbf{e}_{\mathrm{pr}} = \alpha \mathbf{c}\widehat{\mathbf{x}}, \tag{7.1}$$

where α represents the emission reduction rate (assumed as 10%), $\mathbf{c}$ represents the direct emission intensity, and $\mathbf{x}$ represents total output.

The CO_2 emissions (that should be reduced in the target year) induced by the consumption of final goods and services $\mathbf{e}_{\mathrm{cm}}$ is formulated as Eq. (7.2):

$$\mathbf{e}_{\mathbf{cm}} = \alpha\left[\mathbf{c}(\mathbf{I}-\mathbf{A})^{-1}\widehat{\mathbf{f}}\right], \tag{7.2}$$

where $(\mathbf{I} - \mathbf{A})^{-1}$ represents the Leontief inverse matrix, $\widehat{\mathbf{f}}$ denotes the diagonalized matrix of total final demand. The regions with a higher level of final demand-based emissions (higher final demand consumption) should therefore put more effort in emission reduction and be allocated with fewer initial emission allowances.

Several reference approaches that consider other effort-sharing principles have also been proposed. The CO_2 emissions that should be reduced following the equality principle, $\mathbf{e}_{\mathbf{pp}}$, is quantified based on the population proportions of all regions. It is formulated as

$$\mathbf{e}_{\mathbf{pp}} = \mathbf{p}\left(\iota'\mathbf{e}_{\mathbf{pr}}\right), \tag{7.3}$$

where $\mathbf{p}$ represents the adjusted population share of each region (weighted average of the log inverse population value) according to the right-to-development principle. The regions with the larger population should therefore be allocated more initial emission allowances.

The CO_2 emissions that should be reduced following the capacity principle, $\mathbf{e}_{\mathbf{ap}}$, is based on the ability of each region to pay and bear the financial obligations associated with CO_2 emissions. It is formulated as

$$\mathbf{e}_{\mathbf{ap}} = \mathbf{v}\left(\iota'\mathbf{e}_{\mathbf{pr}}\right), \tag{7.4}$$

where $\mathbf{v}$ represents the Gross Regional Product (GRP) share of each region. The regions with a higher level of GRP have more capacity for further emission reduction and should therefore be allocated fewer emission allowances.

The CO_2 emissions that should be reduced following the cost-effectiveness principle, $\mathbf{e}_{\mathbf{co}}$, is based on the MACs of each region. It is formulated as

$$\mathbf{e}_{\mathbf{co}} = \mathbf{m}\left(\iota'\mathbf{e}_{\mathbf{pr}}\right), \tag{7.5}$$

Where $\mathbf{m}$ represents the adjusted MAC indicator of each region (weighted average of the log inverse MAC value) according to the cost-optimal principle. The regions with the lowest marginal abatement costs have the most potential for further emission reduction and should therefore be allocated fewer emission allowances.

2.2 Data Sources

All data sources are listed in Table 7.2.

1. Input-output table

Table 7.2 Data sources

Variable	Description	Sources
$\widehat{\mathbf{x}}$; $\widehat{\mathbf{f}}$; **A**; **v**	Total production vector; Final demand vector; Intermediate input matrix; Value added vector	2012 China 30-region input-output table, CEADs, Mi et al. (2018)
c	Direct emission intensity vector	CEADs, Shan et al. (2017), Shan et al. (2018)
p	Population of each province	China Statistical Yearbook 2013
m	Marginal abatement cost coefficient of each province	Zhou et al. (2013)

Source: Authors

Production ($\widehat{\mathbf{x}}$), final demand ($\widehat{\mathbf{f}}$), input coefficient (**A**), and value-added (**v**) data were obtained from an input-output table. We used the "2012 China 30-region input-output table" issued by Carbon Emission Accounts and Datasets for Emerging Economies [1] (Mi et al. 2018).

2. Direct emission intensity (**c**)

   We used the "2012 China 30-region emission inventories by sectoral approach" data issued by Carbon Emission Accounts and Datasets for Emerging Economies, except for Tibet (Shan et al. 2018). For Tibet, we used the data from Shan et al. (2017).
3. Population (**p**)

   We utilized provincial data from the 2013 China Statistical Yearbook.
4. The relative marginal abatement cost (**m**)

   We used the MAC curve of total CO_2 emissions in each region as estimated by Zhou et al. (2013).

3 Results

3.1 Emission Allowance Allocations According to the Five Approaches

Table 7.3 lists the results of regional cap calculations. The total emissions of the Chinese electricity supply sector in 2020 were 4233.5 Mt-CO2. Thus, with a 10% emission reduction goal, the total initial allowances would be 3810.2 Mt-CO2. The gray cells in the table represent the three provinces with the most and least allocated allowances.

The region Inner Mongolia, Jiangsu, and Shandong are large emitters under the energy consumption-based approach, with 376.1, 335.1, and 322.9 Mt-CO2

[1] https://www.ceads.net/

Table 7.3 Emission allowance allocation by five approaches (Unit: Mt-CO_2)

Region	*(1) Energy consumption-based*	*(2) Final consumption-based*	*(3) Right to development*	*(4) Ability to pay*	*(5) Cost-optimal*
Beijing	23.8	14.2	10.0	-1.6	7.1
Tianjin	58.7	55.9	47.1	35.4	47.4
Hebei	210.7	211.7	223.0	222.4	224.0
Shanxi	203.6	217.2	212.1	215.4	216.6
Inner Mongolia	376.1	402.4	402.2	397.4	406.8
Liaoning	159.8	157.3	164.3	159.5	165.7
Jilin	93.3	92.5	88.5	89.8	87.5
Heilongjiang	99.5	101.0	96.7	99.1	98.1
Shanghai	57.9	51.9	48.4	37.0	47.7
Jiangsu	335.1	323.2	361.5	350.4	363.0
Zhejiang	199.6	194.9	209.5	201.5	208.3
Anhui	158.0	164.5	163.6	166.3	163.1
Fujian	106.2	106.7	104.1	101.1	103.6
Jiangxi	60.9	58.7	54.6	58.5	51.5
Shandong	322.9	318.2	348.9	342.2	351.7
Henan	226.7	231.3	241.9	241.8	241.0
Hubei	118.3	115.3	119.3	119.0	118.7
Hunan	82.2	81.4	79.8	80.6	77.9
Guangdong	228.4	218.9	244.3	236.5	241.6
Guangxi	68.9	69.2	63.6	67.6	58.5
Hainan	13.6	13.1	-4.9	4.7	-7.7
Chongqing	33.5	26.8	22.3	24.7	23.9
Sichuan	54.4	53.0	49.7	50.9	44.9
Guizhou	89.5	94.5	85.2	93.1	92.6
Yunnan	62.3	63.4	56.2	62.1	57.6
Shaanxi	84.9	82.1	80.4	82.0	79.1
Gansu	73.2	75.6	65.8	74.3	66.4
Qinghai	9.6	9.5	-11.2	0.1	-13.8
Ningxia	85.1	89.7	73.2	82.9	77.7
Xinjiang	113.7	116.3	110.2	115.5	109.5
Total	3810.2	3810.2	3810.2	3810.2	3810.2

Source: Authors

allowances being allocated to them initially before emission trading. Inner Mongolia is one of the largest coal-producing regions (Shanxi, similarly) in China, with a significant portion of its economy dependent on the coal industry While Jiangsu and Shandong both hold a diverse range of industries, including petrochemicals, textiles, and food processing. On the other hand, the region Qinghai, Hainan, and Beijing are the small emitters under the energy consumption-based approach, with 9.6, 13.6, and 23.8 Mt-CO2 allowances being allocated to them initially before emission trading. As a relatively small region in terms of population and economic activity, Qinghai has been known for its high-altitude plateau landscapes and rich mineral resources, while Hainan is an island province known for its tropical climate and beautiful beaches. In contrast to them, Beijing is the capital city of China with a population of over 21 million people. The total emission number of Beijing (Shanghai, similarly) was small due to its relatively small geographical size and holding of fewer industrial economic activities.

Regarding the difference in initial allocation under different effort-sharing principles, it would be better shown in the next section. However, according to Table 7.3, it can be observed that the differences in large emitters' initial allowance among different effort-sharing principles would be occupying a relatively smaller share. Thus, which principle the initial allocation is based on would largely affect small emitter regions.

3.2 Adjustment of Emission Allowance Allocations in All Regions

The regional caps were assumed to be set using the default (1) energy consumption-based approach and actual emission calculations for all regions to be conducted using the (2) final consumption-based, (3) right to develop, (4) ability to pay, and (5) cost-optimal approaches.

The inter-regional CO2 trade allowances were calculated as the difference between the values in Column (1) and those in Columns (2)–(5). The results are presented in Table 7.4. Column (1) of Table 7.4 shows the default capping values of the regions and is identical to column (1) of Table 7.3.

The regions that would be able to sell a larger portion of their initial emission allowances (dark grey cells) under most effort-sharing principles include Shanxi and Inner Mongolia. They are also the larger emitter regions with a lower population density (see Column (3)), a certain level of financial capability to achieve the emission reduction target (relatively higher GRP per capita, see Column (4)) and a larger emission reduction potential (lower marginal abatement cost, see Column (5)). However, the emissions within these two regions were also induced by the final demand of other regions. Therefore, after adjustment, these regions would be able to sell some of their initial emission allowances.

The regions that would need to purchase a larger proportion of emission allowances (light gray cells) under most effort-sharing principles include Beijing, Shanghai, and Chongqing. They are larger consumers that induced the emissions generated outside the region, at the same time, with smaller emission reduction potential (higher marginal abatement cost).

The regions that would largely change their strategies (from selling to purchasing, or vice versa) under different effort-sharing principles include Jiangsu, Shandong, and Guangdong. Regarding Jiangsu and Shandong, with the financial capability (relatively higher GRP per capita, see Column (4)) and larger emission reduction potential (lower marginal abatement costs, see Column (5)), the two regions may sell some emission allowances under the *Ability to Pay* and *Cost-optimal* effort-sharing

Table 7.4 Emission allowance changes compared to the energy consumption-based approach (Unit: Mt CO_2)

Region	*(1) Energy consumption -based*	*(2) Final consumption -based*	*(3) Right to development*	*(4) Ability to pay*	*(5) Cost-optimal*
Beijing	23.8	–9.6	–13.8	–25.4	–16.6
Tianjin	58.7	–2.8	–11.5	–23.3	–11.2
Hebei	210.7	1.0	12.3	11.7	13.3
Shanxi	203.6	13.6	8.5	11.9	13.0
Inner Mongolia	376.1	26.2	26.1	21.3	30.7
Liaoning	159.8	–2.5	4.5	–0.4	5.8
Jilin	93.3	–0.8	–4.9	–3.5	–5.8
Heilongjiang	99.5	1.6	–2.8	–0.4	–1.3
Shanghai	57.9	–6.0	–9.4	–20.9	–10.1
Jiangsu	335.1	–11.9	26.5	15.3	27.9
Zhejiang	199.6	–4.7	9.9	1.9	8.7
Anhui	158.0	6.5	5.6	8.3	5.1
Fujian	106.2	0.5	–2.1	–5.1	–2.6
Jiangxi	60.9	–2.2	–6.4	–2.5	–9.5
Shandong	322.9	–4.8	26.0	19.3	28.8
Henan	226.7	4.6	15.2	15.1	14.3
Hubei	118.3	–3.0	1.1	0.8	0.5
Hunan	82.2	–0.8	–2.4	–1.6	–4.2
Guangdong	228.4	–9.5	15.9	8.1	13.2
Guangxi	68.9	0.2	–5.3	–1.3	–10.5
Hainan	13.6	–0.5	–18.5	–8.9	–21.3
Chongqing	33.5	–6.7	–11.2	–8.7	–9.5
Sichuan	54.4	–1.4	–4.6	–3.4	–9.5
Guizhou	89.5	5.0	–4.3	3.6	3.1
Yunnan	62.3	1.1	–6.1	–0.2	–4.7
Shaanxi	84.9	–2.8	–4.5	–2.9	–5.7
Gansu	73.2	2.5	–7.4	1.1	–6.8
Qinghai	9.6	–0.1	–20.8	–9.6	–23.4
Ningxia	85.1	4.6	–11.9	–2.2	–7.4
Xinjiang	113.7	2.7	–3.5	1.8	–4.2
Total	3810.2	0.0	0.0	0.0	0.0

Notes: e.g., a negative value –11.9 of Jiangsu under *Final Consumption-based* means, if the emission accounting is conducted based on a *Final consumption-based* approach, Jiangsu would need to purchase another 11.9 Mt emission allowances

Source: Authors

principle. However, they would both need to purchase emission allowances under the *final consumption-based* principle, as they induced emissions generated in other regions. Liaoning will also change the strategies of initial allowances under different effort-sharing principles, with selling under the *Right-to-development* and *Cost-optimal* effort-sharing principle, while purchasing under the *final consumption-based* and *Ability to Pay* effort-sharing principle.

4 Discussion

Despite the fact that the Chinese national ETS currently only applies to the electricity generation sector, it is anticipated to be gradually expanded to include a total of eight sectors: petrochemical, chemical, building materials, steel, nonferrous metals, paper, and domestic aviation. The adoption of mitigation strategies in these industries among sub-regions and within these eight sectors may differ significantly. For example, the potential to utilize and access renewable energies differ among sub-regions and sectors; the same is true for green hydrogen. Such disparities may act as a barrier to efficient and equitable decarbonization transitions. Regional characteristics should be considered when determining the initial allocations of emission trading in China. In particular, access to renewable energy and the potential of bonus allowances for the utilization of green hydrogen would be the next stage of investigation following this study.

5 Conclusion

This study used the *energy consumption-based* (i.e., emissions generated from energy use) and *final consumption-based* (i.e., emissions induced by the consumption of final goods and services) approaches, in conjunction with three reference approaches (*right to development*, *ability to pay*, and *cost-optimal approaches*) to provide a reference framework for further adjustment of initial allowance allocations.

An appropriate emission allocation within an ETS can expedite the transition towards electricity-saving technologies and lifestyles in energy-intensive regions. Jiangsu, Beijing, and Guangdong would be required to purchase larger proportions of emission allowances if emission allowances were calculated using a *final consumption-based* approach; these additional mitigation costs could inversely change the behavior of electricity consumers, ultimately resulting in a wider range of emission reductions.

Furthermore, appropriate emission allocations could support the improvement of emission intensity in regions that generate a large amount of electricity. Shanxi and Inner Mongolia would receive a greater number of emission allowances if the calculation of emission allowances were based on the *final consumption-based* approach; this revenue could be used to phase out coal and introduce renewable energy sources.

Regarding the *cost-optimal* reference approach, the regions with low MACs could initially share emission reduction efforts. However, the marginal abatement cost of a region would vary dynamically with time and cumulative emission reductions. These dynamic changes and the constraints on the potential for emission reduction in each region were not addressed in this study.

Acknowledgements This work was supported by JSPS KAKENHI Grant Numbers JP19K12459372 and JP21H04941.

References

van den Berg NJ, van Soest HL, Hof AF, den Elzen MG, van Vuuren DP, Chen W, Drouet L, Emmerling J, Fujimori S, Höhne N, Kõberle AC (2019) Implications of various effort-sharing approaches for national carbon budgets and emission pathways. Clim Change:1–18

Höhne N, den Elzen M, Escalante D (2014) Regional GHG reduction targets based on effort sharing: a comparison of studies. Clim Policy 14(1):122–147. https://doi.org/10.1080/14693062.2014.849452

International Carbon Action Partnership (ICAP) (2021) ETS detailed information: China. https://icapcarbonaction.com/en/ets-map.

Mi Z, Meng J, Zheng H, Shan Y, Wei YM, Guan D (2018) A multi-regional input-output table mapping China's economic outputs and interdependencies in 2012. Sci Data 5:180155

Shan Y, Guan D, Zheng H, Ou J, Li Y, Meng J, Mi Z, Liu Z, Zhang Q (2018) China CO2 emission accounts 1997–2015. Sci Data 5:170201. https://doi.org/10.1038/sdata.2017.201

Shan Y, Zheng H, Guan D, Li C, Mi Z, Meng J, Schroeder H, Ma J, Ma Z (2017) Energy consumption and CO2 emissions in Tibet and its cities in 2014. Earth's Future 5(8):854–864

Zhou P, Zhang L, Zhou DQ, Xia WJ (2013) Modeling economic performance of interprovincial CO 2 emission reduction quota trading in China. Appl Energy 112(2013):1518–1528. https://doi.org/10.1016/j.apenergy.2013.04.013

Chapter 8
Recycling Resources from End-of-Life Vehicles in China

Yang Li and Yiyi Ju

Contents

Keywords End-of-life vehicles · Weibull distribution · Recyclable resource potential · Urban mine

Abbreviation

ELV End-of-life vehicles

1 Introduction

With the rapid increase in the number of vehicles in China, the number of end-of-life vehicles is expected to skyrocket. To manage this crisis, construction of a recycling system is currently being explored. The accumulation of end-of-life vehicles, which

Y. Li
Zhongnan University of Economics and Law, Hubei, China
e-mail: z0004698@zuel.edu.cn

Y. Ju (✉)
Waseda University, Tokyo, Japan
e-mail: juyiyi@aoni.waseda.jp

K. Fujikawa (ed.), *Empirical Research on Environmental Policies in China*,
https://doi.org/10.1007/978-981-99-5957-0_8

contain a wide variety of metal materials and chemicals, are now becoming clearly visible "urban mines". The increase of end-of-life vehicles and the increased demand for resources due to economic growth are making the recycling of China's automotive resources an urgent and challenging task.

It takes about ten years for a car to reach the end of its life after being shipped. We, therefore, a highly reliable method for estimating the number of end-of-life vehicles and the recyclable resource potential of those vehicles to promote appropriate recycle plans. The purpose of this paper is to estimate the number of end-of-life vehicles generated using Weibull distribution, and then to estimate the amount of resource and material contained in end-of-life vehicles. This paper is positioned as basic research for building a resource recycling system in China.

2 Prediction of the Number of New Registrations of Civil Vehicles in China

It is necessary to estimate the number of new vehicle registrations in the future to project the future number of end-of-life vehicles. There are a variety of opinions, optimistic and pessimistic, regarding the future trend of the automobile market in China. However, the automobile market in China's inland urban and rural areas is expected to expand further as incomes rise. In this paper, it is assumed that China is still at the initial stage of motorization, and the automobile market in China will continue to expand until 2035 driven by the expansion of the inland market. Therefore, this study first estimates the future domestic demand using an approximate curve. An approximate curve is a trend line of market expansion estimated based on observed data, and future values can be predicted by extending this trend line into the future. Polynomial equations are often used to forecast future growth of durable goods since polynomial equations can take future slowdown in the growth into account [1]. The estimated equation is shown as Eq. (8.1):

$$P(x) = \begin{array}{ccc} -0.032x^2 & +2.123x & -1.844 \\ (-1.78) & (5.55) & (-1.05) \end{array} \tag{8.1}$$

The sample period is 20 years from 2002 to 2021. P stands for the number of new registrations of vehicles, and x stands for year that begins with 1 in 2002 and ends with 20 in 2021.The t-values of the coefficients of the primary and secondary

[1]Li et al. (2022).

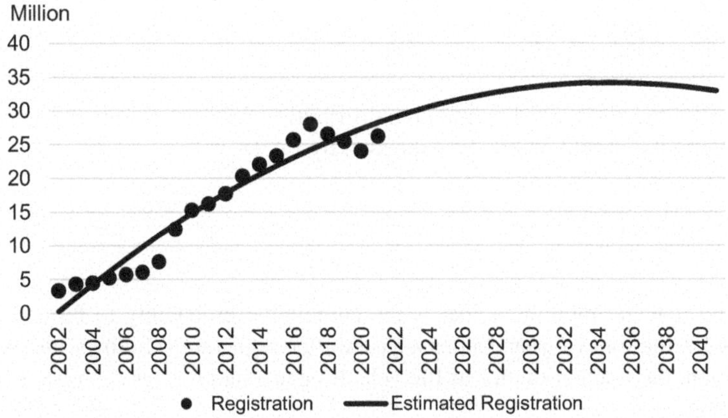

Fig. 8.1 Prediction of the number of new vehicle registrations in China. Source: Authors' estimation

explanatory variables are significant at a significance level of 5% and R-square is 0.939, which is sufficiently high.

Figure 8.1 shows the estimation results. If we estimate the number of new registrations of civil vehicles in China based on Eq. (8.1), there will be 40.7 million vehicles in 2035.

3 Number of End-of-life Vehicles in China

3.1 Weibull Distribution

We will now estimate the number of end-of-life vehicles based on the number of new registrations and owned vehicles using the Weibull distribution. The Weibull distribution is a probability distribution proposed by W. Weibull to statistically describe the intensity of an object. When the variable t stands for a "time-to-failure", the Weibull distribution gives a distribution for which the cumulative failure rate is proportional to a power of time. The probability density function of the Weibull distribution is expressed as Eq. (8.2) (Weibull (1951)).

$$f(t) = \frac{m}{\eta}\left(\frac{t}{\eta}\right)^{m-1} exp\left[-\left(\frac{t}{\eta}\right)^{m}\right] \quad (0 \leq t, 0 < m, 0 < \eta) \tag{8.2}$$

where m is the shape parameter and η is the scale parameter of the Weibull distribution. The left-sided cumulative distribution function (cumulative failure probability) of the Weibull distribution can be expressed by Eq. (8.3).

$$W(t) = 1 - exp\left[-\left(\frac{t}{\eta}\right)^m\right] \tag{8.3}$$

Incidentally, the right-sided cumulative distribution function (survival probability) of the Weibull distribution can be expressed by Eq. (8.4).

$$R(t) = exp\left[-\left(\frac{t}{\eta}\right)^m\right] \tag{8.4}$$

Assuming that the ratio $W_{t,\ k}$ that is the cumulative probability to be end-of-life vehicles by the end of year t after k years after registration follows the Weibull distribution, the number of end-of-life vehicles generated in year t can be estimated by Eq. (8.5).

$$\widehat{B}_t = \sum_{k=1}^{n} P_{t,k}\, W_{t,k} - \sum_{k=1}^{n} P_{t-1,k}\, W_{t-1,k} \tag{8.5}$$

$\widehat{B}_t$:	The number of estimated end-of-life vehicles in year t
$P_{t,\ k}$:	The number of new vehicle registrations in year $t - k$
$W_{t,\ k}$:	Cumulative scrap rate for vehicles of age k in year t
t:	Target year of calculation
n:	Maximum years used
k:	Age of vehicles

3.2 *Results of Estimating the Number of End-of-life Vehicles*

It is considered that the number of vehicles owned and the number of end-of-life vehicles will further rapidly grow along with the increase in automobile sales. it is necessary to estimate the shape parameter m and the scale parameter η to estimate the number of end-of-life vehicles using the Weibull distribution. We use $m = 5.14$, $\eta = 11.26$ following Wu et al. (2022). Figure 8.2 shows the estimated number of end-of-life vehicle generation from 2022 to 2035.

The obtained result show that the number of end-of-life vehicles on the Chinese mainland was forecasted to be 26.5 million vehicles in 2030 and 31.5 million vehicles in 2035. Additionally, the annual mean increase rate from 2010 to 2035 was forecasted to be approximately 11%.

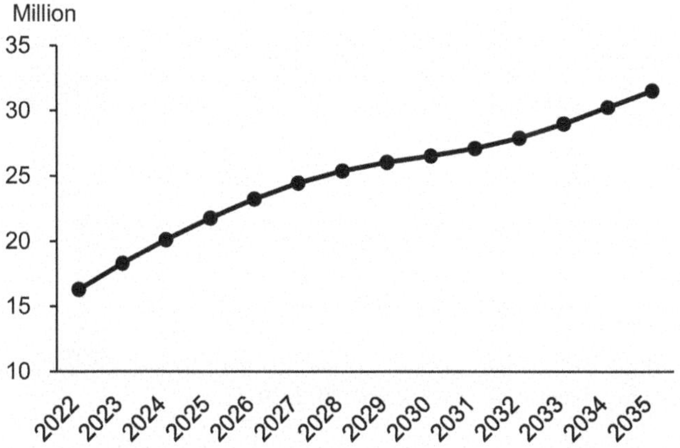

Fig. 8.2 Estimation of the number of end-of-life vehicles. Source: Authors' estimation

While estimating the number of scrapped vehicles, it is necessary to consider the changes in the scrapping system. This is because that changes in the scrapping system may affect the number of end-of-life vehicles. In China, regulations such as the "End-of-Life Vehicle Standards (1997)," the "Notice on Regulations to Adjust End-of-Life Vehicle Standards (2000)," and the "Measures for Managing the Recovery of Scrapped Vehicles (2001)," among others, require that vehicles be scrapped when they reach a certain age or mileage, depending on vehicle and usage types. However, the restriction on vehicle age was abolished in the "Standard Rules for Mandatory Disposal of Mobile Vehicles" enacted on May 1, 2013.Under the current law, passenger vehicles are to be scrapped when their mileages reach 600,000 km, regardless of their usage years. Conventionally, the general mileage of vehicles in China was approximately 20,000 km per year, and the period of use was 15 years. The abolition of the scrappage age limit may extend the number of years a vehicle can be used, and the actual number of years of use is thought to be 15 years or more. Therefore, the impact of such changes in laws and regulations should be taken into account as an error factor in the future prediction of the number of end-of-life vehicles. However, as this factor was not considered in this study, changes in laws and regulations should be included in the estimation model or not is left as a future issue.

4 Estimating the Recyclable Resource Potential of End-of-life Vehicles

As the number of vehicles owned is increasing year by year, it is expected that the number of end-of-life vehicles will also rapidly increase in the future. Looking from a different perspective, it is forecasted that the end-of-life vehicles will lead to "urban mine." Given this situation, we estimated the number of end-of-life vehicles

Table 8.1 The weight share of recyclable resources in ELVs

Recyclable resources	Passenger vehicles	Buses	Trucks
Steel	71.9%	72.0%	80.6%
Non-ferrous metal	7.7%	7.0%	5.2%
Plastic	11.8%	7.7%	3.3%
Rubber	2.1%	4.8%	5.8%
Glass	1.3%	3.2%	1.1%
Liquid	0.9%	1.1%	0.9%
Other materials	4.3%	4.2%	3.1%
Average unit weight of ELVs	1.30 ton	2.45 ton	4.50 ton
The share of vehicle type in ELVs	42.84%	30.03%	27.13%

Source: China Automobile Dealers Association (2012, 2021)

generated in China, which is expected to increase rapidly in the future. Based on this, we estimated the recyclable resource potential of end-of-life vehicles. However, in this study, we estimated this potential under the following two conditions.

First, we estimate the recyclable resource potential based on our assumption that the end-of-life vehicles estimated in the previous section were all recovered through official distribution route. As the number of exported Chinese used vehicles is extremely small, and if vehicles are collected through official distribution routes, almost all end-of-life vehicles should be processed domestically. However, the current situation is that the scrapping procedures in the official distribution route is complicated, and a higher profit can be obtained by selling them to unlicensed dismantlers on the black market. Therefore, a large number of end-of-life vehicles are recovered and dismantled through unauthorized distribution routes. As it is very difficult to grasp the informal market, we will exclude it in this study.

Secondly, as the weight of recycled resources that can be recovered from end-of-life vehicles varies greatly depending on the type of vehicle, we have divided end-of-life vehicles into passenger vehicles, buses, and trucks, and performed an estimate for each type in this study. On the other hand, as there is no major difference in the recycled resource ratio between new energy vehicles and conventional fuel vehicles [2], we do not distinguish between new energy vehicles and conventional fuel vehicles when estimating the recyclable resource potential of end-of-life vehicles in this study.

Below, we firstly estimate the recyclable resource potential for each vehicle type, and then we describe our comprehensive estimation results for the recyclable resource potential included in end-of-life vehicles. Table 8.1 shows the constitution rate of recycled resources and ELVs that can be recovered from end-of-life vehicles by vehicle type.

For year t, the ratio of i vehicle types among the $\widehat{B}_t$ estimated end-of-life vehicles is S_i, and the j types of recycled resources per i vehicle type are recovered, of which total weight is C_{ij}. The ratio of each resource is I_{ij}. The recyclable resource potential

[2]China Automotive Technology & Research Center (2017).

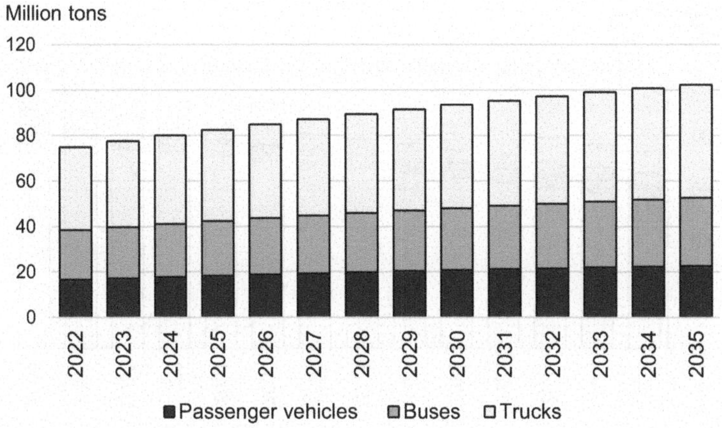

Fig. 8.3 Recyclable resources from ELVs by type of vehicles from 2022 to 2035. Source: own elaboration based on China Automobile Dealers Association (2021)

by material type and vehicle type $\widehat{q}_{ij}(t)$, and their total amount $\widehat{Q}_{ij}(t)$ can be estimated by Eqs. (8.6) and (8.7), respectively.

$$\widehat{q}_{ij}(t) = I_{ij}S_iC_{ij}\widehat{B}_t \quad (8.6)$$

$$\widehat{Q}(t) = \sum_{j=1}^{h}\sum_{i=1}^{d} I_{ij}S_iC_{ij}\widehat{B}_t \quad (8.7)$$

$\widehat{q}_{ij}(t)$:	Recycled of resource *j* from vehicle type *j* in year *t*
$\widehat{Q}(t)$:	Total recycled resources for end-of-life vehicles in year *t*
$\widehat{B}_t$:	Estimated number of end-of-life vehicles in year *t*
I_{ij}:	Ratio of recovered recycled resources *j* from vehicle type *i*
S_i:	Ratio of vehicle type *i* among total number of end-of-life vehicles
C_{ij}:	Total weight of recycled resource *j* included per one unit of vehicle type *i*
t:	Target year of calculation
h:	Type of recovered recycled resource
d:	Vehicle type

The recyclable resource potential for each vehicle type estimated based on the estimation results of end-of-life passenger vehicles is shown in Fig. 8.3. The recyclable resource potential recovered from end-of-life passenger vehicles, buses and trucks is expected to reach 22.7 million tons, 29.9 million tons, 49.7 million tons, respectively by 2035.

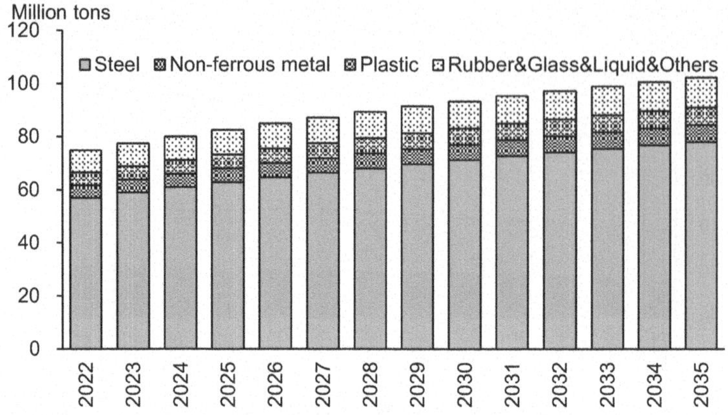

Fig. 8.4 Recyclable resources from ELVs by resource from 2022 to 2035. Source: own elaboration based on China Automobile Dealers Association (2012, 2021)

The estimated recyclable resource potential by resource type is shown in Fig. 8.4. The results show that of the 93.4 million tons recycled resources that will be recovered from the end-of-life vehicles in 2030, there will be 71.1 million tons of iron, 5.9 million tons of nonferrous metals, 6.1 million tons of plastics, 4.4 million tons of rubber, and 1.6 million tons of glass. In 2035, the potential recycled resources recoverable from end-of-life vehicles is expected to be 102.3 million tons, the breakdown of which is 77.9 million tons of iron, 6.4 million tons of nonferrous metals, 6.6 million tons of plastics, 4.8 million tons of rubber, 1.8 million tons of glass, and 1.0 million tons of oils.

5 Conclusion

It is necessary to estimate the number of end-of-life vehicles that will be generated in future to establish a vehicle recycling system. This paper estimated the number of end-of-life vehicles from 2022 to 2035 using a Weibull distribution based on time-series data on the number of newly registered and owned vehicles in China, and estimated the potential of China's end-of-life vehicle recycling resources. From the estimation result, it is predicted that the quantity of recycled resources that can be recovered from end-of-life vehicles will be 93.4 million tons in 2030 and will reach 102.3 million tons in 2035.

Here, we discuss some future issues related to this paper. Firstly, while estimating the number of end-of-life vehicles in this paper, we estimated the probability of vehicle scrappage by grouping all vehicles together. However, to obtain more accurate estimated numbers, it is necessary to estimate the scrap rate by vehicle type. Further, in this paper, estimation of vehicle scrapping probability does not consider possible changes in the government scrapping policy, and the resource recovery quantity was estimated based on very rough assumptions of technological progress.

As already mentioned in Sect. 3.2, standards for vehicle scrapping in China have recently changed in the previous regulations, for Mandatory Disposal of Mobile Vehicles" enacted in 2013 that private passenger vehicles must be scrapped when their mileage reaches 600,000 km regardless of the vehicle age. It should be noted that such changes in laws and regulations can be disturbance factors in the prediction of the future number of end-of-life vehicles. Before this new regulation was enacted, the average annual mileage of passenger cars in China was approximately 20,000 km, so that although the mileage restriction did not apply after 15 years, many cars were scrapped after 15 years of vehicle age. However, the elimination of the vehicle age-based scrapping regulation may increase the number of years a vehicle can be used. In the future, we would like to consider incorporating such regulatory changes into the model.

Acknowledgements This work was supported by JSPS KAKENHI Grant Numbers JP19K12459372 and JP21H04941.

References

China Automobile Dealers Association (2012, 2021) China Auto Market Almanac. China Business Press. (in Chinese)

China Automotive Technology & Research Center (2017) *Automotive industry green development report*; Posts & Telecom Press. ISBN 978-7-115-47175-8. (in Chinese)

Li Y, Liu Y, Chen Y, Huang S, Ju Y (2022) Projection of end-of-life vehicle population and recyclable metal resources: provincial-level gaps in China. Sustain Prod Consum 31:818–827. https://doi.org/10.1016/j.spc.2022.03.034

Weibull W (1951) A statistical distribution function of wide applicability. J Appl Mech 18(3): 293–297. https://doi.org/10.1115/1.401033

Wu C, Li Y, Zhang Y, Liu Y, Huang S, Ju Y (2022) Long-term estimation of plastic material resources from end-of-life vehicles in China: a scenario analysis considering multiple industry standards. J Mater Cycles Waste Manag. https://doi.org/10.1007/s10163-022-01380-2

Chapter 9
Carbon Leakage in Carbon Taxes and Emissions Trading Scheme Taking China as an Example

Hikari Ban and Kiyoshi Fujikawa

Contents

Keywords Computable general equilibrium model · GTAP model · Carbon tax · Emissions trading scheme · Carbon leakage · China

H. Ban (✉)
Kobe Gakuin University, Kobe, Japan
e-mail: ban@eb.kobegakuin.ac.jp

K. Fujikawa
Aichi Gakuin University, Nagoya, Japan
e-mail: fujikawa@dpc.agu.ac.jp

K. Fujikawa (ed.), *Empirical Research on Environmental Policies in China*,
https://doi.org/10.1007/978-981-99-5957-0_9

Abbreviations

CGE	Computable general equilibrium
CO_2	Carbon dioxide
ETS	Emission trading scheme
EU	Europe union
EV	Equivalent variation
GDP	Gross domestic product
GTAP	Global Trade Analysis Project
ICT	Industry-wise carbon tax

1 Introduction

In September 2020, Chinese President Xi Jinping formally declared in a video speech at the United Nations Summit on Biodiversity that the country aims to reach peak carbon (CO_2) emissions by 2030 and CO_2 neutrality by 2060. In recent years, various measures have been implemented to achieve these goals. One of these measures is emissions trading. In 2013 and 2014, two provinces and five cities, including Shanghai, Beijing, and Guangdong, began emission trading pilot projects, with Fujian and Sichuan provinces joining later in 2018. July 2021 saw the launch of China's National ETS (Emissions Trading Scheme) for its electric power sector, the world's largest in terms of emissions. In the first year after its launch, the cumulative trading volume was about 194 million tons of CO_2 and the total trading value was about 8.492 billion RMB (1.25 billion USD) (Yang 2022). In addition, during the 14th Five-Year Plan period (2021–2025), the scheme will be expanded to include petrochemicals, chemicals, construction materials, steel, non-ferrous metals, paper-making, and aviation as soon as possible[1].

This study uses a multiregional, multisectoral computable general equilibrium (CGE) model to estimate how the economic and environmental effects of an industry-wise carbon tax and the domestic emissions trading in China differ for each sector. We also consider carbon leakage in non-target sectors in China and carbon leakage outside China.

2 Prior Research

The design of emissions trading schemes has been studied from various perspectives. Considering the EU-ETS (European Union Emissions Trading System), Dijkstra et al. (2011) used a two-country, three-sector partial equilibrium model

[1] For more information on China's emissions trading scheme, see ICAP (2022).

incorporating international emissions trading. The authors demonstrated that in theory, total economic welfare is maximized when all sectors are subject to emissions trading.

Several analyses have been presented based on a dynamic CGE model with China's ETS in mind. Mu et al. (2018) indicated that an ETS covering the entire industry minimizes its negative impact on GDP and that an ETS with only eight sectors, which the Chinese government has proposed for emissions trading, would significantly negatively affect GDP. They stated that as many energy-intensive industries as possible should be included in the ETS system. Jiang et al. (2022) suggested that an ETS that includes electricity, chemicals, non-ferrous metals, crude oil, transportation, and construction sectors would have the least negative impact on GDP, assuming that electricity, chemicals, and non-ferrous metals sectors participate in the ETS. This is because chemicals and non-ferrous metals are capital-intensive, so the substitution of energy for capital would increase investment and mitigate the negative effect on GDP. In addition, Lin and Jia (2017) observed that ETS sector coverage was insignificant in the negative effect on GDP. However, Lin and Jia (2020) revised the original conclusion by the same authors to argue that, in general, having more sectors covered by the ETS mitigates the negative effect on GDP and lowers the ETS price.

Thus, while there may be some consensus that the economic burden is minimized when all sectors are subject to the ETS, there is not always agreement on which sectors should be included when the ETS is only partially implemented. Therefore, to clarify the factors that cause differences in the economic effects of different ETS target industries, this study assumes 16 hypothetical scenarios with different ETS target industries and analyzes their effects on China's domestic and international environment and economy.

3 Model and Data

3.1 GTAP-E Model

This study uses Truong's version of the GTAP-E model (Truong 2007)[2], which is one of the CGE models that include economic, energy, and environmental interdependence. The model consists of the following equilibria: goods market equilibrium, factor market equilibrium, zero-profit condition, and emissions trading market equilibrium. Savings and investment are in equilibrium globally, with global investment being equal to global savings. A global bank is assumed to invest the

[2]The GTAP-E model is based on the GTAP model, a CGE model developed by the Global Trade Analysis Project (GTAP) at Purdue University. It is a model that links energy and the environment to economic activity. For details of the model, see Hertel (1997), Burniaux and Truong (2002). The model used here is available in the study by Truong (2007).

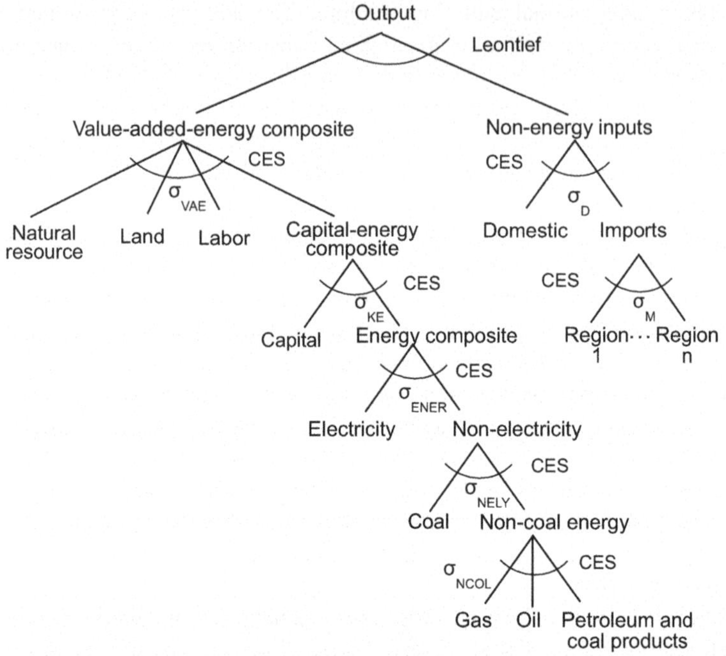

Fig. 9.1 Structure of production function in GTAP-E. Source: prepared by the author based on the study by Burniaux and Truong (2002). Note: Leontief in the figure indicates that the aggregate is a linear function; CES indicates that the aggregate is a CES function. In addition, σ represents the elasticity of substitution of the CES function

savings from each country, and the global bank decides to invest in each country so that the expected rate of change in the rate of return on each country's investment is equal. If emissions trading is assumed, the equilibrium equation for the emissions trading market is added to the model.

It is necessary to understand the production function to interpret the simulation results. Figure 9.1 depicts the production function in the GTAP-E model. At the top, value-added-energy composites and non-energy inputs produce output with fixed coefficients. The rest are functions of the constant elasticity of substitution (CES), where the value-added-energy composites comprise natural resources, land, labor, and capital-energy composites. Furthermore, energy composites consist of gas, petroleum/coal products, coal, and electricity in three stages, as illustrated in Fig. 9.1.

The Armington assumption[3] is also imposed, and an import composite is formed by selecting an import source that minimizes costs and comparing the price of the domestic good to that of the import composite to select the optimal combination.

[3]Under the Armington assumption, goods in the same industry are not in a relationship of perfect substitution but have a certain elasticity of substitution depending on the producing country; see Armington (1969).

Table 9.1 Regional integration of the GTAP data

No.	Code	Description	No.	Code	Description
1	OCE	Oceania	8	US	United States of America
2	Chn	China & Hong Kong	9	Latin	Latin America
3	Jpn	Japan	10	EUEFTA	EU & European Free Trade Association
4	Kor	Korea	11	Rus	Russia
5	ASEAN	Association of Southeast Asian Nations	12	CAC	Central Asia & Caucasus
6	Ind	India	13	MENA	Middle East and North Africa
7	Can	Canada	14	ROW	Rest of the world

Source: Prepared by the author based on GTAP-E Database 10A

3.2 *Data*

For analysis, we used the GTAP-E Database 10A, which corresponds to the global economy in 2014. We integrated it into 14 regions and 29 sectors, as presented in Tables 9.1 and 9.2. In the regional classification, the United Kingdom is included in EUEFTA. CAC consists of Kazakhstan, Kyrgyzstan, Tajikistan, other former Soviet Union countries (Turkmenistan and Uzbekistan aggregated), Armenia, Azerbaijan, and Georgia.

Next, we examine the parameter values critical for interpreting the simulation results and each sector's CO_2 emissions and energy mix.

For the production structure in Fig. 9.1, the price elasticity of demand for non-coal energy composites consisting of gas, oil, and petroleum/coal products σ_{NCOL}; the elasticity of non-electricity composites consisting of coal and non-coal energy composites σ_{NELY}; the elasticity of energy composites consisting of electricity and non-electricity composites σ_{ENER}; and the elasticity of capital-energy composites consisting of capital input and energy composites σ_{KE} by industry are summarized in Table 9.3. The most notable feature is that the values of elasticity of energy and capital substitution regarding coal, gas, and petroleum/coal products are set to zero, or not substitutable with other inputs as is shown in Table 9.3.

The elasticity σ_{VAE} of value-added-energy composites consisting of natural resources, land, labor, and capital-energy composites is 1.26 for most sectors, including the electricity sector. However, some sectors, such as gas, have different values in different countries[4]. Armington elasticity takes large values for products such as gas and oil, for which the differences between countries are minor[5].

Table 9.4 presents the CO_2 emissions of China's top 15 industries and their CO_2 and energy intensity per unit of output. The top three CO_2 emitters are electricity,

[4]For example, the σ_{VAE} for the gas sector in China, Japan, and the United States are 0.15, 0.01, and 0.33, respectively.

[5]σ_D and σ_M for Chinese gas and oil are 11.74 and 5.20, and 32.92 and 10.40, respectively.

Table 9.2 Departmental integration in the GTAP data

No.	Code	Description	No.	Code	Description
1	Agri	Grains and Crops	16	NMM	Non-metallic mineral products
2	Lvstc	Livestock	17	I_S	Iron/steel
3	Forest	Forestry	18	Auto	Automobile and parts
4	Fish	Fishing	19	T_Equ	Transportation equipment
5	Coal	Mining of Coal	20	E_Equ	Electrical/electronic equipment
6	Oil	Extraction of crude oil	21	M_Equ	Machin equipment
7	Gas	Extraction of natural gas and distribution	22	O_Mnf	Other manufacturing
8	PPP	Papermaking/publishing	23	Water	Water
9	P_C	Petroleum/coal products	24	Const	Construction
10	Ely	Electricity	25	Trade	Trade
11	Mining	Other mining	26	A_Trs	Air transport service
12	Food	Processed food	27	W_Trs	Water transport service
13	Tex	Textiles/wearing apparel	28	L_Trs	Land transport service
14	NFM	Non-ferrous metals	29	Svc	Other Services
15	Chm	Chemical products			

Source: Prepared by the author based on GTAP-E Database 10A

Table 9.3 Elasticity of energy and capital substitution in the GTAP-E model

	σ_{KE}	σ_{ENER}	σ_{NELY}	σ_{NCOL}
Coal	0	0	0	0
Gas	0	0	0	0
Oil	0	0	0	0
P_C	0	0	0	0
Electricity	0.5	0	0.5	1
Other Industries	0.5	1	0.5	1

Source: Prepared by the author based on GTAP-E Database 10A

non-metallic mineral products, and iron/steel. These sectors constitute 70.8% of total industry emissions and 87.3% of the eight sectors expected to be included in China's ETS.

Table 9.5 presents the energy composition of China's major sectors. A high share of coal demand characterizes the electricity, non-metallic mineral products, and coal sectors. The three transportation sectors (land, water, and air) are characterized by a high ratio of demand for petroleum/coal products. The non-ferrous metals sector is characterized by a high demand ratio of electricity.

Table 9.4 Industry characteristics of China's top 15 CO_2-emitting industries

	CO_2 emissions M ton	Share %	CO_2 intensity Tons/1 million USD	Energy intensity toe/1 million USD
Ely	4,061.7	53.9	8,206.3	2,298.0
NMM	716.9	9.5	838.3	274.5
I_S	558.1	7.4	423.0	181.7
L_Trs	459.0	6.1	501.2	187.9
Chm	383.6	5.1	226.7	153.3
Svc	209.1	2.8	33.9	18.6
P_C	196.9	2.6	318.4	1,929.0
Coal	129.3	1.7	635.8	210.7
Food	102.7	1.4	65.4	27.1
W_Trs	84.9	1.1	608.5	214.9
A_Trs	81.4	1.1	771.4	272.9
O_Mnf	75.8	1.0	36.0	26.8
NFM	73.2	1.0	92.4	85.2
Mining	46.3	0.6	125.7	75.9
PPP	44.5	0.6	126.0	69.5

Source: Prepared by the author based on GTAP-E Database 10A

Table 9.5 Energy (Mtoe) composition of China's top 15 CO_2 emitting industries: %

	Coal	Oil	Gas	P_C	Ely
Ely	92.0	0.0	1.9	2.6	3.6
NMM	73.0	0.0	2.5	10.6	13.9
I_S	36.7	0.0	1.6	36.5	25.2
L_Trs	2.0	0.0	13.2	81.0	3.8
Chm	26.2	1.5	8.3	36.9	27.1
Svc	30.8	0.1	11.1	16.6	41.5
P_C	36.9	42.1	0.8	18.9	1.3
Coal	72.3	0.0	0.0	7.3	20.4
Food	58.4	0.0	4.5	4.1	33.0
W_Trs	0.4	0.0	0.4	98.9	0.3
A_Trs	0.5	0.0	0.1	98.6	0.7
O_Mnf	26.5	0.3	4.0	14.2	55.0
NFM	19.9	0.0	3.7	10.3	66.1
Mining	27.0	0.0	2.2	22.1	48.7
PPP	44.8	0.0	2.9	3.2	49.1

Source: Prepared by the author based on GTAP-E Database 10A

4 Simulation Scenario

Table 9.6 indicates each simulation scenario, where ICT represents industry-wise carbon tax and ETS denotes Emissions Trading Scheme. The numbers in the latter half of the scenario name indicate the number of industries covered by the scheme. The values in the Table 9.6 are reduction rates (%), which are set to reduce the total CO_2 emissions of the target sectors by 1223.2 million tons. In other words, for ICT01, only the electric power sector will be reduced by 30.1%. For ICT08, each of the eight sectors will be reduced by 20%. The same applies to ETS, with ETS02 reducing only the electricity and non-metallic mineral products sectors by 25.6%.

When all industries are targeted, carbon taxes and emissions trading are two sides of the same coin. For carbon tax, if the carbon tax rate is determined first, the amount of CO_2 reduction is determined by the price effect of the demand function. Furthermore, in emissions trading, if the amount of CO_2 reduction is determined first and then emissions trading is conducted, the price of carbon is determined as a result of market equilibrium.

However, the assumptions made for the industry-wise carbon tax simulations in this study are somewhat different from those for a regular carbon tax. The industry-wise carbon tax simulations (ICT01 to ICT08) calculate the industry-wise carbon tax needed to achieve the industry's CO_2 reduction target. In other words, the carbon tax rate differs for each industry. Moreover, the emissions trading simulations (ETS02 to

Table 9.6 CO_2 reductions for each simulation scenario (%)

	Ely	NMM	NFM	I_S	Chm	PPP	P_C	A_Trs	Others
ICT01	−30.1								
ICT02	−25.6	−25.6							
ICT03	−25.2	−25.2	−25.2						
ICT04	−22.6	−22.6	−22.6	−22.6					
ICT05	−21.1	−21.1	−21.1	−21.1	−21.1				
ICT06	−21.0	−21.0	−21.0	−21.0	−21.0	−21.0			
ICT07	−20.3	−20.3	−20.3	−20.3	−20.3	−20.3	−20.3		
ICT08	−20.0	−20.0	−20.0	−20.0	−20.0	−20.0	−20.0	−20.0	
ETS02	−25.6	−25.6							
ETS03	−25.2	−25.2	−25.2						
ETS04	−22.6	−22.6	−22.6	−22.6					
ETS05	−21.1	−21.1	−21.1	−21.1	−21.1				
ETS06	−21.0	−21.0	−21.0	−21.0	−21.0	−21.0			
ETS07	−20.3	−20.3	−20.3	−20.3	−20.3	−20.3	−20.3		
ETS08	−20.0	−20.0	−20.0	−20.0	−20.0	−20.0	−20.0	−20.0	
ETS29	−20.0	−20.0	−20.0	−20.0	−20.0	−20.0	−20.0	−20.0	0.0

Note: In the table, reduction rates are shown to one decimal place, but in reality, impacts are set to six decimal places. A value of 0.0 for Others in ETS29 indicates that CO_2 emissions in each sector other than the eight sectors are constrained to maintain the status quo.
Source: Prepared by the author based on GTAP-E Database 10A

ETS29) are the same as usual, and calculations are done for the carbon price necessary to achieve the CO_2 reduction target under the condition that emissions trading is permitted.

5 Carbon Pricing and Emissions Trading

5.1 Carbon Pricing

Table 9.7 presents the industry-wise carbon tax and carbon prices in emissions trading for each scenario. In this study carbon price means not the price of ton-CO_2 but the price of ton-C. The first comparison of carbon prices for each sector in the ICT08 scenario reveals that petroleum/coal products has the highest carbon price, followed by air transportation. Conversely, papermaking/publishing has the lowest carbon price, followed by non-metallic mineral products. The main factors that define the carbon price are the value of the energy-related elasticity of substitution (Table 9.3) and the energy composition of each sector (Table 9.5). The GTAP-E model in this study assumes no capital or energy substitution in petroleum/coal products, resulting in higher carbon prices for these products. In the air transportation sector, petroleum/coal products constitute 98.6% of the energy input, so there is no room for substitution with low-carbon energy sources such as gas, resulting in a high carbon price. In papermaking/publishing, coal and electricity constitute a relatively large share of the energy input, and thus a lower carbon price is possible through coal-to-electricity substitution. In the non-metallic mineral products industry, the share of coal is large, and the shares of electricity and petroleum/coal products are also relatively large, so substitution away from coal will lead to a lower carbon price.

Table 9.7 Carbon price (US$/ton-C)

ICT										
	Ely	NMM	NFM	I_S	Chm	PPP	P_C	A_Trs	ETS	
ICT01	26.0									
ICT02	20.7	19.7							ETS02	20.5
ICT03	20.4	19.2	27.8						ETS03	20.3
ICT04	17.9	16.2	23.8	31.8					ETS04	18.5
ICT05	16.5	14.7	21.7	28.5	26.4				ETS05	17.4
ICT06	16.4	14.6	21.5	28.2	26.1	12.4			ETS06	17.2
ICT07	16.7	12.0	16.1	17.7	18.7	11.4	486.8		ETS07	17.0
ICT08	16.4	11.8	15.9	17.5	18.5	11.2	467.0	68.9	ETS08	17.0
									ETS29	12.6

Source: Prepared by the author based on GTAP-E Database 10A

Table 9.8 Domestic emissions trading (M ton-CO_2)

	ETS02	ETS03	ETS04	ETS05
Ely	−5.8	−2.7	29.0	42.6
NMM	5.8	7.0	15.3	18.9
NFM		−4.3	−3.3	−2.9
I_S			−41.0	−36.3
Chm				−22.3
	ETS06	**ETS07**	**ETS08**	**ETS29**
Ely	39.7	60.9	69.1	−98.2
NMM	18.6	23.3	24.8	−3.8
NFM	−2.9	−2.3	−2.2	−4.6
I_S	−36.1	−31.5	−30.3	−44.3
Chm	−22.2	−19.3	−18.5	−28.9
PPP	2.9	3.1	3.2	1.1
P_C		−34.0	−33.3	−31.7
A_Trs			−12.9	−13.7
Others				224.0

Source: Prepared by the author based on GTAP-E Database 10A

In the ICT scenario described in this chapter, the carbon price decreases as the number of sectors increases. This is because the abatement burden for each sector decreases[6].

Next, comparing ETS02 to ETS29, which have the same total amount of reduction obligations for target sectors, we see that the carbon price also decreases as the number of target sectors increases. Incidentally, the average of each industry-wise carbon tax in the simulation is calculated as follows: ICT02: \$20.5, ICT03: \$20.3, ICT04: \$19.2, ICT05: \$18.1, ICT06: \$17.9, ICT07: \$31.7, ICT08: \$31.3[7]. Thus, comparing the industry-wise carbon tax and the emissions trading scenarios, we can say that the carbon price is lower for emissions trading than for the carbon tax as the number of sectors covered increases.

5.2 *Emissions Trading*

Table 9.8 indicates the change in CO_2 emissions and the amount of emissions trading under each scenario. A positive number in the Table 9.8 means that CO_2 allowances are sold, while a negative number means that CO_2 allowances are purchased.

[6]The exception is that the carbon price in the electricity sector increases from ICT06 to ICT07. This is because petroleum/coal products is included in the carbon tax in ICT07, and the price of petroleum/coal products increases while the price of coal decreases, which requires a higher carbon tax on electricity, which has a higher share of coal and a lower share of petroleum/coal products.

[7]The average of each industry-wise carbon tax was calculated by dividing the total carbon tax revenue by the total emissions of the relevant sector.

Few industries become sellers, with only the following three industries taking on the role. Papermaking/publishing is a seller in the ETS06 and later scenarios, non-metallic mineral products is a seller in all besides the EST29 scenario, and electric power is a seller in the ETS04 through ETS08 scenarios.

In the ETS29 scenario, all industries subject to emissions trading are buyer industries except for papermaking/publishing, and all other industries are sellers. The top five seller industries with the largest supply of allowances are coal mining (49.0 million tons), service (44.7 million tons), land transportation (32.8 million tons), food (22.9 million tons), and gas extraction (13.8 million tons).

Theoretically, industries with lower marginal abatement costs become sellers, and sectors with higher marginal abatement costs become buyers. In this sense, the rate of the industry-wise carbon tax provides an indication of the difference in marginal abatement costs by industry. The industry-wise carbon tax in Table 9.7 for ICT06 indicates that the tax rate are lower in the papermaking/publishing, non-metallic mineral products, and electricity sectors in this order. These sectors are as a result the sellers in the ETS06 scenario of emissions trading.

6 Macro Impact

6.1 Macroeconomic Impact

Table 9.9 lists the macroeconomic impacts on GDP, equivalent variation (EV), wages, and capital rental rates. In the case of the industry-wise carbon tax simulation,

Table 9.9 Impact on the Chinese macroeconomy

	GDP (%)	EV, Mill. USD	Wage rate (%)	Capital rental rate (%)
ICT01	−0.29	−45,904.0	−1.28	−2.26
ICT02	−0.26	−42,208.7	−1.18	−2.06
ICT03	−0.26	−42,232.4	−1.21	−2.08
ICT04	−0.26	−42,030.4	−1.30	−2.12
ICT05	−0.26	−41,331.6	−1.37	−2.17
ICT06	−0.26	−41,110.5	−1.37	−2.16
ICT07	−0.47	−62,903.2	−2.05	−3.43
ICT08	−0.46	−61,404.5	−2.05	−3.43
ETS02	−0.26	−42,210.2	−1.18	−2.06
ETS03	−0.26	−42,184.6	−1.20	−2.07
ETS04	−0.26	−41,312.0	−1.25	−2.07
ETS05	−0.25	−40,609.0	−1.30	−2.11
ETS06	−0.25	−40,364.4	−1.30	−2.10
ETS07	−0.25	−40,649.0	−1.32	−2.14
ETS08	−0.25	−40,429.8	−1.33	−2.16
ETS29	−0.22	−35,273.7	−1.16	−2.02

Source: Prepared by the author based on GTAP-E Database 10A

the impact on GDP and EV is not significantly different in the ICT01 to ICT06 industry-wise carbon scenarios. However, the impact on GDP and EV is slightly more significant in ICT07 and ICT08 when petroleum/coal products is included in the emissions reductions. This is because the GTAP-E model assumes that capital and energy are mutually not substitutable in petroleum/coal products. Hence, the economic impact is more prominent when the petroleum/coal products industry is included in the reductions. The impact on the wage and capital rental rates does not differ for the ICT01 to ICT06 scenarios and is slightly larger for ICT07 and ICT08.

Furthermore, the impact of domestic emissions trading on GDP and EV is largest in ETS02 and smallest in ETS29. The economic burden is not as significant in ETS07 as in the ICT scenario for the carbon tax, even if petroleum/coal products is included in the reduction targets. The impact on the wage and capital rental rates does not decrease as much in the ETS scenario as in the ICT scenario, even if petroleum/coal products is included in the abatement. The wage and capital rental rates decrease is the smallest in ETS29.

The effects of emissions trading are estimated by comparing the ICT carbon tax scenarios with the ETS emissions trading scenarios. The impacts on GDP, EV, wage rates, and capital rental rates tend to be mitigated by the introduction of emissions trading[8]. The degree of mitigation is greater in the scenarios in which petroleum/coal products is targeted for reductions.

6.2 *Carbon Leakage*

The total CO_2 emission reductions for all industries subject to reduced emissions in the simulation scenarios in this study are set to 1,223.2 million tons in each case. The difference in the overall reductions in China under each simulation scenario results from differences in carbon leakage to industries and households not obligated to reduce their CO_2 emissions (Table 9.10). The industry-specific impacts in the next section reveal that CO_2 reduction policies affect CO_2 emissions through various pathways. However, a crucial channel of CO_2 leakage to non-target sectors is through the effect of lower coal prices. CO_2 reductions in the power sector reduce coal demand, lowering coal prices. Lower coal prices increase demand for coal in non-target sectors, increasing CO_2 emissions.

Consider the ICT scenario. CO_2 emissions from coal mining, a non-target industry, are reduced in all scenarios, as indicated in Tables 9.14 and 9.15. However, in scenarios ICT01 to ICT03, the substitution effect for coal in non-target industries exceeds the effect of the reduction in coal production, resulting in carbon leakage to the non-target sectors. In scenarios following ICT04, energy-intensive industries

[8]The exceptions are the EV of ICT02 and ETS02, with ETS02 being slightly worse. This is since the two sectors are electricity and non-metallic mineral products, with no difference in industry-wise carbon taxes, which are both estimated around \$20.

Table 9.10 CO_2 changes and carbon leakage in China

	CO_2 changes				Carbon leakage				
	Domestic total		Industry total		Non-target industries	Household	Overseas	Domestic	Overseas
	M ton	%	M ton	%	M ton	M ton	M ton	%	%
ICT01	−1,164.3	−14.4	−1,179.2	−15.7	44.1	14.9	−7.6	4.8	−0.7
ICT02	−1,190.4	−14.7	−1,204.2	−16.0	19.1	13.7	−3.7	2.7	−0.3
ICT03	−1,196.2	−14.8	−1,209.9	−16.1	13.4	13.7	−2.7	2.2	−0.2
ICT04	−1,213.9	−15.0	−1,227.1	−16.3	−3.8	13.2	4.8	0.8	0.4
ICT05	−1,228.4	−15.2	−1,241.7	−16.5	−18.4	13.3	6.6	−0.4	0.5
ICT06	−1,231.2	−15.2	−1,244.6	−16.5	−21.3	13.3	6.6	−0.7	0.5
ICT07	−1,340.6	−16.6	−1,334.5	−17.7	−111.3	−6.1	38.2	−9.6	2.8
ICT08	−1,331.3	−16.5	−1,325.8	−17.6	−102.5	−5.5	42.8	−8.8	3.2
ETS02	−1,190.7	−14.7	−1,204.4	−16.0	18.8	13.7	−3.5	2.7	−0.3
ETS03	−1,196.4	−14.8	−1,210.0	−16.1	13.2	13.6	−2.8	2.2	−0.2
ETS04	−1,212.1	−15.0	−1,225.5	−16.3	−2.3	13.4	1.8	0.9	0.2
ETS05	−1,226.3	−15.2	−1,239.9	−16.5	−16.6	13.6	3.3	−0.2	0.3
ETS06	−1,229.3	−15.2	−1,242.9	−16.5	−19.6	13.6	3.3	−0.5	0.3
ETS07	−1,233.0	−15.2	−1,245.9	−16.5	−22.7	12.9	4.6	−0.8	0.4
ETS08	−1,233.0	−15.2	−1,245.9	−16.5	−22.7	12.9	6.4	−0.8	0.5
ETS29	−1,217.2	−15.0	−1,223.2	−16.2	0.0	6.1	7.3	0.5	0.6

Note: The domestic carbon leakage rate is the ratio of the total change in CO_2 emissions from non-target sectors and households divided by the reductions in non-target sectors. The global carbon leakage rate is the ratio of the change in CO_2 emissions from outside China divided by the reduction in domestic CO_2 emissions.

Source: Prepared by the author based on GTAP-E Database 10A

such as iron/steel, and non-metallic mineral products are targeted for reduction. Thus, the decrease in coal production has a greater effect, meaning that carbon leakage does not occur. Instead, negative carbon leakage occurs, whereby CO_2 emissions are reduced even in non-target industries. Furthermore, negative carbon leakage is maximized in ICT07, where petroleum/coal products is targeted for reduction. In the ICT08 scenario, the air transportation sector, which had previously contributed to the negative carbon leakage to non-target industries, is included in the reduction target, resulting in lesser negative carbon leakage.

In the ETS scenarios, CO_2 emissions from the coal mining sector decrease. However, in ETS02 and ETS03, the effect of substituting coal for energy exceeds that of reducing coal production, resulting in carbon leakage. However, in the scenarios after ETS04, the effect of the reduction in coal production becomes larger, resulting in negative carbon leakage. In the ETS scenario, however, the negative carbon leakage is not as large as in the ICT scenario, even though petroleum/coal products is subject to reductions. The reason is that the increase in petroleum/coal product prices is lower than that in the ICT scenario. Like the industry-wise carbon tax, carbon leakage changes as the number of industries covered by the tax expands. Naturally, in the ETS29 scenario, carbon leakage is limited to the household portion and is very small.

Let us also look at carbon leakage to other countries. A sign of carbon leakage to foreign countries is the opposite direction of that for the domestic non-target industry sector. In the low-target industry scenarios, exports from energy-intensive goods industries, such as chemicals and iron/steel in China, grow due to lower production factor prices. This will result in a decrease in production in those industries abroad, leading to negative carbon leakage. As the scope of emission reductions in China increases, exports from China will decrease and production abroad will increase, leading to carbon leakage. By country, there is no significant change in CO_2 emissions. However, there is a relatively noticeable negative carbon leakage of approximately −0.1% for Japan, Korea, Canada, and CAC in the ICT01 and ICT02 scenarios and a 0.3% positive carbon leakage for EUEFTA and MENA in the ICT07 and ICT08 scenarios.

China dominates the global CO_2 emissions trends, with the largest reductions being −4.3% (−1302.4 million tons) in the ICT07 scenario in ICT and −4.1% (−1226.6 million tons) in the ETS08 scenario in the ETS.

7 Impact by Industry

7.1 Impact on the Energy Sector

Changes in energy prices and production are summarized in Table 9.11 and 9.12. The CO_2 reduction policy increases the price of electricity and decreases its production. The coal sector experiences lower prices and production due to weaker demand for coal from the power sector. Gas, which has a higher cost share of electricity, will

Table 9.11 Impact on China's domestic energy prices (%)

Number of sectors		1	2	3	4	5	6	7	8	29
Coal	ICT	−2.0	−2.2	−2.3	−2.4	−2.5	−2.5	−3.4	−3.4	–
	ETS	–	−2.2	−2.3	−2.3	−2.4	−2.5	−2.5	−2.5	−1.2
Oil	ICT	0.4	0.2	0.1	−0.2	−0.4	−0.4	−4.8	−4.8	–
	ETS	–	0.2	0.1	−0.1	−0.2	−0.2	−0.4	−0.4	−0.6
Gas	ICT	2.7	1.7	1.6	1.0	0.3	0.2	1.7	1.5	–
	ETS	–	1.7	1.6	1.1	0.5	0.5	0.4	0.3	1.9
P_C	ICT	0.5	0.3	0.2	0.0	−0.1	−0.1	18.3	17.5	–
	ETS	–	0.3	0.2	0.1	0.0	0.0	0.7	0.6	0.4
Ely	ICT	18.9	15.2	14.9	13.0	11.9	11.8	12.4	12.1	–
	ETS	–	15.1	14.8	13.5	12.6	12.5	12.4	12.3	9.4

Source: Prepared by the author based on GTAP-E Database 10A

increase in price and decrease in production. These phenomena occur in all scenarios.

Petroleum/coal products constitute a small share of electricity, so the impact of electricity price increases is small. However, a carbon tax on petroleum/coal products would increase their prices greatly. In the first three scenarios, the production of petroleum/coal products increases due to the substitution effect from electricity. However, when sectors with relatively high-cost shares of petroleum/coal products, such as iron/steel and chemicals, are subject to reductions, the demand for and production of petroleum/coal products decrease.

The impact on oil price caused by electricity price increase is relatively small because of its low-cost share of electricity, and oil prices decline in the scenarios with more than four sector targets. Oil production tends to increase slightly due to substitution for electricity but decreases in ICT07 and ICT08 when the production of petroleum/coal products declines dramatically.

7.2 *Impact on Prices, Imports and Exports, and Production*

Table 9.13 presents the effects on prices, imports/exports, and production in the target sectors under the eight sector scenarios. Typically, CO_2 reduction policies cause higher prices, lower exports, higher imports, and lower production in target sectors. For imports, the overall reduction in production due to CO_2-reducing policies lowers import demand, but the import-increasing effect of higher prices for domestic goods prevails. These trends are generally observed in all scenarios, not just the eight sector scenarios.

However, the opposite is observed in the papermaking/publishing sector. This is because electricity and petroleum/coal products account for a smaller share of total costs in the industry. In addition, the impact of lower prices on wage and capital rental rates outweighs the impact of higher prices for these products due to CO_2

Table 9.12 Impact on China's domestic energy production (%)

Number of sectors		1	2	3	4	5	6	7	8	29
Coal	ICT	−17.1	−17.5	−17.5	−17.7	−17.9	−17.9	−22.9	−22.6	–
	ETS	–	−17.5	−17.5	−17.7	−17.9	−17.9	−18.1	−18.1	−17.0
Oil	ICT	0.9	0.8	0.8	0.5	0.4	0.4	−3.8	−3.7	–
	ETS	–	0.8	0.8	0.6	0.5	0.5	0.4	0.4	−0.3
Gas	ICT	−14.5	−12.9	−13.0	−12.8	−15.6	−15.8	−19.0	−18.7	–
	ETS	–	−12.9	−13.0	−13.2	−15.7	−15.9	−17.3	−17.2	−44.2
P_C	ICT	1.0	0.6	0.4	−0.7	−1.1	−1.1	−16.2	−16.0	–
	ETS	–	0.5	0.4	−0.2	−0.6	−0.6	−1.3	−1.4	−2.2
Ely	ICT	−13.1	−10.6	−10.4	−8.9	−8.1	−8.0	−7.4	−7.3	–
	ETS	–	−10.5	−10.3	−9.3	−8.7	−8.6	−8.5	−8.4	−6.5

Source: Prepared by the author based on GTAP-E Database 10A

Table 9.13 Impacts on prices, imports/exports, and production by industry in China (8 sector scenarios) %

	Price		Import		Export		Production	
	ICT08	ETS08	ICT08	ETS08	ICT08	ETS08	ICT08	ETS08
PPP	−0.4	−0.0	−2.2	−0.8	4.0	1.5	0.4	0.1
P_C	17.5	0.6	12.3	−0.8	−45.0	−1.8	−16.0	−1.4
Ely	12.1	12.3	15.2	14.4	−38.2	−39.4	−7.3	−8.4
NFM	0.1	0.4	−0.4	0.5	1.5	−1.0	0.8	0.1
Chm	1.1	0.4	1.2	0.1	−4.0	−0.8	−1.3	−0.4
NMM	0.9	1.3	−0.4	1.6	−2.0	−3.9	−3.3	−2.2
I_S	1.7	0.9	1.5	0.6	−5.8	−3.0	−2.7	−1.6
A_Trs	6.9	0.7	7.5	0.2	−15.1	−1.1	−7.3	−0.7

Source: Prepared by the author based on GTAP-E Database 10A

reduction policies[9]. Similarly, non-ferrous metals, which have a lower cost share of electricity and petroleum/coal products, experience smaller price increases under the ICT08 scenario and lower prices relative to their overseas counterparts. This leads to higher exports and lower imports[10]. Papermaking/publishing and non-ferrous metals production will increase due to export growth. Furthermore, the increased production of non-ferrous metals is also linked to the increased production of electrical/electronic equipment[11].

Imports also decrease in non-metallic mineral products in the ICT08 scenario[12]. This may be related to a reduction (−4.3%) in production in the construction sector. In the ETS08 scenario, the price increase rate of petroleum/coal products is suppressed due to emissions trading, which has a smaller effect on the increase in imports. Imports of petroleum/coal products decrease due to a drop in demand caused by general economic stagnation.

Carbon reduction policies also negatively impact output in non-target sectors. However, production in agriculture, food, textiles/wearing apparel, electrical/electronic equipment, and other manufacturing industries increases in both ICT08 and

[9]The top five shares of total costs for electricity and petroleum/coal products are as follows. Gas 39.6%, petroleum/coal products 29.2%, air transportation 21.6%, water transportation 17.2%, and land transportation 13.6%. Papermaking/publishing and non-ferrous metals constituted 3.2% and 5.6%, respectively.

[10]Outside of China, wage rates and capital rental rates increase and most output prices rise. The price changes for non-ferrous metals are 0.4% for Japan, 0.3% for Korea, 0.3% for ASEAN, 0.5% for the U.S., and 0.4% for the EU and EFTA in the ICT08 scenario.

[11]The cost share of non-ferrous metals in the cost of electrical/electronic equipment is as high as 10.0%.

[12]The decrease in imports of non-ferrous metals and non-metallic mineral products, which is observed in eight sector scenarios, is only observed in the scenarios where petroleum/coal products is targeted. The increase in production of non-ferrous metals occurs only in ICT05, 06, 07, and 08, where demand from exports and the electrical/electronic equipment sector is high.

Table 9.14 Change in CO_2 emissions by major industries in the ICT scenario (Unit: M ton-CO_2)

	ICT01	ICT02	ICT03	ICT04	ICT05	ICT06	ICT07	ICT08
Ely	−1,223.2	−1,039.7	−1,024.0	−918.4	−857.6	−851.1	−823.3	−812.3
NMM	17.5	−183.5	−180.8	−162.1	−151.4	−150.2	−145.3	−143.4
NFM	6.0	5.2	−18.5	−16.6	−15.5	−15.3	−14.8	−14.6
Iron	15.7	13.0	12.8	−126.2	−117.8	−116.9	−113.1	−111.6
Chm	14.6	13.4	13.5	13.2	−81.0	−80.4	−77.8	−76.7
PPP	3.6	3.2	3.2	3.0	2.9	−9.3	−9.0	−8.9
P_C	1.6	1.1	0.8	−1.1	−1.7	−1.7	−39.9	−39.4
A_Trs	0.2	0.2	0.2	0.3	0.4	0.4	−5.6	−16.3
W_Trs	−0.3	−0.2	−0.2	−0.2	−0.2	−0.2	−5.7	−5.6
L_Trs	−5.0	−4.2	−4.2	−4.0	−3.7	−3.7	−36.3	−35.2
Coal	−41.8	−40.9	−40.9	−40.7	−40.7	−40.7	−51.7	−50.9
Other Industries	31.9	28.4	28.1	25.7	24.6	24.5	−11.9	−10.9
Non-target industries	44.1	19.1	13.4	−3.8	−18.4	−21.3	−111.3	−102.5

Source: Prepared by the author based on GTAP-E Database 10A

ETS08 scenarios due to export growth. Although carbon reduction policies typically reduce output in non-target industries, production in industries that gain international competitiveness from lower factor prices may increase[13].

7.3 CO_2 Emissions by Industry

Tables 9.14 and 9.15 indicate the change in CO_2 emissions for the major industries. CO_2 emissions in target sectors decrease regardless of the existence of emissions trading. Paper products/publishing and non-ferrous metals see decreased CO_2 due to energy substitution, despite the expansion of production.

CO_2 emissions from coal and water/land transportation decrease in all scenarios. The decrease in coal emissions is particularly pronounced. CO_2 emissions increase in most non-target industries except coal, water/land transportation, especially in the case the number of target sectors is small. However, in ICT07 and ICT08, CO_2 emissions decrease in many sectors as the rise in the price of petroleum/coal products gives grate negative impact on the production of many sectors.

[13] The growth rates of production in ICT08 and ETS08 are 0.6% and 0.5% for agriculture, 0.1% and 0.2% for food, 2.4% and 1.5% for textiles/wearing apparel, 3.0% and 1.8% for electrical/electronic equipment, and 0.1% and 0.1% for other manufacturing, respectively.

Table 9.15 Change in CO_2 emissions by major industries under ETS scenario (Unit: M ton-CO_2)

	ETS02	ETS03	ETS04	ETS05	ETS06	ETS07	ETS08	ETS29
Ely	−1,033.9	−1,021.3	−947.4	−900.2	−890.8	−884.2	−881.5	−714.2
NMM	−189.3	−187.7	−177.4	−170.3	−168.9	−168.6	−168.2	−139.6
NFM	5.2	−14.2	−13.2	−12.6	−12.4	−12.5	−12.4	−10.1
Iron	12.9	12.8	−85.2	−81.5	−80.8	−81.6	−81.4	−67.4
Chm	13.4	13.4	13.2	−58.7	−58.1	−58.4	−58.2	−47.8
PPP	3.2	3.2	3.1	3.0	−12.2	−12.1	−12.1	−10.0
P_C	1.0	0.9	−0.3	−0.8	−0.7	−5.9	−6.1	−7.7
A_Trs	0.2	0.2	0.3	0.3	0.3	0.1	−3.4	−2.6
W_Trs	−0.2	−0.2	−0.2	−0.2	−0.2	−0.4	−0.4	−2.7
L_Trs	−4.2	−4.2	−3.9	−3.7	−3.7	−4.9	−4.9	−32.8
Coal	−40.9	−40.9	−40.8	−40.9	−40.9	−41.3	−41.2	−49.0
Other Industries	28.2	28.1	26.5	25.6	25.5	23.9	23.8	−139.5
Non-target industries	18.8	13.2	−2.3	−16.6	−19.6	−22.7	−22.7	0.0

Source: Prepared by the author based on GTAP-E Database 10A

8 Discussion

Similar to the results of previous studies, this study also shows that industry-wide emissions trading has the lowest economic impact. Even loose constraints on CO_2 emissions seem to lower the overall economic impact by inducing sellers to become buyers in the emissions trading market.

The simulations in this chapter were analyzed by including each industry targeted for CO_2 reduction one at a time. As such, the results may differ if the order of the additions was different. However, the economic burden of including petroleum/coal products in the reduction targets is likely to remain high. This result is not necessarily consistent with those of previous studies, which may be due to differences in the assumption of the elasticity of substitution of energy as an input good and the elasticity of substitution of imported goods among countries.

The GTAP-E Database 10A used in this study is a database that incorporates the international interdependence of each country's trade. Therefore, the simulation results reflect the change in international trades affected by China's CO_2 reduction policies. For example, under the carbon reduction policy, China's production in most sectors declines, but some sectors experience increased production, such as electrical/electronic equipment and textiles wearing apparel. This is because the carbon reduction policy lowers wage and capital rental rates and increases exports. We also found that as China's energy-intensive goods sectors become target sectors for emission reductions, overseas production in these sectors increases, creating carbon leakage in a global context.

There are many issues with this study. First, it does not consider the initial allocation method, which is critical to the design of the ETS. The GTAP-E model is a fee-based allocation in which sectors with reduction obligations pay an emissions surcharge for the entire amount of CO_2 they emit, and other allocation methods are not considered.

Second, although our study has chosen between a carbon tax and emissions trading, one possible approach to eliminate domestic carbon leakage is to divide industries into those that trade emissions and those that do not and to add a carbon tax to those that do not. Under these circumstances, in addition to the issue of how to distinguish between the two, the ratio of emission allowances allocated to the emissions trading industry and the carbon tax industry would also pose an issue[14].

9 Conclusion

This study used a CGE analysis to analyze how the economic and environmental effects of an industry-wise carbon tax and domestic emissions trading are affected by the expansion of the sectors covered by the tax and trading in the Chinese economy. The main conclusions of our study are as follows.

Emissions trading covering all industries was shown to have the smallest economic impact. If partial emissions trading is implemented in the eight or so sectors the Chinese government has proposed, it may be preferable to mitigate the macro impacts of emissions trading by excluding petroleum/coal products, which have high marginal abatement costs.

We found that carbon leakage is affected by the expansion of the target sectors. While there is carbon leakage in non-target sectors due to lower coal prices, CO_2 reductions in the electricity sector have significant negative carbon leakage (i.e., lower CO_2 emissions) effects on the coal sector. In turn, CO_2 reductions in the petroleum/coal products sector have significant negative carbon leakage (i.e., lower CO_2 emissions) effects on the transportation sectors and households. When many energy-intensive sectors are non-target sectors, the effects of carbon leakage exceed those of negative carbon leakage, resulting in carbon leakage within China. As the number of industries targeted for emission reductions increases, the latter's effect exceeds the former's, and domestic negative carbon leakage can occur.

Ironically, however, when negative carbon leakage occurs in China, carbon leakage occurs overseas. This is because when China imposes CO_2 emission limits on many energy-intensive goods and production declines, the overseas production of energy-intensive goods increases to take their place. This suggests that in today's highly globalized economy, it is not enough for a country to set domestic CO_2 emission changes as a policy target when implementing carbon reduction policies. International policy coordination that also takes international carbon leakage into account is required.

[14]Studies that are focused on the EU-ETS with this issue in mind include those by Bernard et al. (2004), Böhringer and Rosendahl (2009), and Malueg and Yates (2009).

Acknowledgements This work was supported by JSPS KAKENHI Grant Numbers JP19K12459, JP20K12291, JP21H04941.

References

Armington PS (1969) A theory of demand for products distinguished by place of production. *Int Monet Fund Staff Pap* 16(1):159–178

Bernard A, Vielle M, Viguier L (2004) Modeling the European directive establishing a scheme for greenhouse gas allowance trading and assessing the market power of firms. presented in The 7th Annual Conference on Global Economic Analysis, Washington DC, United States. Retrieved October 12, 2021 from https://www.gtap.agecon.purdue.edu/resources/download/1754.pdf

Böhringer C, Rosendahl KE (2009) Strategic partitioning of emission allowances under the EU emission trading scheme. *Resour Energy Econ* 31(3):182–197. https://doi.org/10.1016/j.reseneeco.2009.04.001

Burniaux J-M & Truong TP (2002). GTAP-E: an energy-environmental version of the GTAP Model. *GTAP Technical Paper, 16*. Retrieved June 10, 2020 from https://www.gtap.agecon.purdue.edu/resources/download/1203.pdf.

Dijkstra BR, Manderson E, Lee T-Y (2011) Extending the sectoral coverage of an international emission trading scheme. *Environ Resour Econ* 50(2):243–266. https://doi.org/10.1007/s10640-011-9470-1

Hertel TW (ed) (1997) *Global trade analysis: modeling and applications*. Cambridge University Press, Cambridge. https://doi.org/10.1017/CBO9781139174688

International Carbon Action Partnership (ICAP) (2022) Emissions trading worldwide: status report 2022. International Carbon Action Partnership. Retrieved May 10, 2022 from https://icapcarbonaction.com/en/publications/emissions-trading-worldwide-2022-icap-status-report

Jiang H-D, Liu L-J, Dong K, Fu Y-W (2022) How will sectoral coverage in the carbon trading system affect the total oil consumption in China? A CGE-based analysis. *Energy Econ* 110: 105996. https://doi.org/10.1016/j.eneco.2022.105996

Lin B, Jia Z (2017) The impact of emission trading scheme (ETS) and the choice of coverage industry in ETS: A case study in China. *Appl Energy* 205:1512–1527. https://doi.org/10.1016/j.apenergy.2017.08.098

Lin B, Jia Z (2020) Does the different sectoral coverage matter? An analysis of China's carbon trading market. *Energy Policy* 137:111164. https://doi.org/10.1016/j.enpol.2019.111164

Malueg DA, Yates AJ (2009) Strategic behavior, private information, and decentralization in the European Union Emissions Trading System. *Environ Resour Econ* 43:413–432. https://doi.org/10.1007/s10640-009-9274-8

Mu Y, Evans S, Wang C, Cai W (2018) How will sectoral coverage affect the efficiency of an emissions trading system? A CGE-based case study of China. *Appl Energy* 227:403–414. https://doi.org/10.1016/j.apenergy.2017.08.072

Truong TP (2007) GTAP-E: an energy-environmental version of the GTAP model with emission trading—USER'S GUIDE. GTAP Resource #2509. Retrieved June 10, 2020 from https://www.gtap.agecon.purdue.edu/resources/res_display.asp?RecordID=2509

Yang Y (2022) China's carbon emission trading market to celebrate 1 year anniversary, China Daily, Global Edition. Retrieved June August 7, 2022 from https://global.chinadaily.com.cn/a/202207/15/WS62d133dda310fd2b29e6ca1b.html

Chapter 10
Environmental Effects of Plastic Waste Recycling: Compilation and Application of Waste Input–Output Table in China

Xi Lu and Kiyoshi Fujikawa

Contents

Keywords Plastic waste · Plastic waste recycling · Waste input–output table · Landfill

1 Introduction

Since 2010, China's rapid industrialization, urbanization, and rising living standards have rapidly increased the amount of discharged waste. In particular, the amount of discharged plastic waste has increased substantially due to the increased consumption of plastic products. Therefore, the Chinese government hopes to reduce the

X. Lu
Hunan University of Finance and Economics, Changsha, Hunan, China
e-mail: luxi@hufe.edu.cn

K. Fujikawa (✉)
Aichi Gakuin University, Nagoya, Aichi, Japan
e-mail: fujikawa@dpc.agu.ac.jp

K. Fujikawa (ed.), *Empirical Research on Environmental Policies in China*,
https://doi.org/10.1007/978-981-99-5957-0_10

amount of waste for final disposal and of natural resources consumed by increasing the collection rate of renewable resources, including plastics. This study reviews the current status of plastic waste recycling in China and estimates how much landfill waste can be reduced by promoting the recycling of the plastic waste.

2 Current Status of Plastic Waste Recycling in China

Domestic consumption of plastics in China continued to increase from 22,640 thousand tons in 2003, reaching a peak of 75,670 thousand tons in 2017; it declined slightly to ~60,000 thousand tons in 2020. However, as shown in Fig. 10.1, the domestic collection rate of the plastics remains at less than 30%, and more than 40,000 thousand tons of the plastic waste is discarded every year without being recycled. The problem with China's plastic waste is that the amount of plastic waste collected does not keep pace with the amount of the plastics consumed.

As shown in Fig. 10.1, prior to 2017, when imports of the plastic waste were restricted, the recycling rate of the plastic waste was calculated using the sum of domestically collected plastic waste plus imported plastic waste as the denominator. Since the imported plastic wastes are for recycling, an increase in the imported plastic waste leads to an increase in the recycling rate of the plastic waste. Indeed, the amount of the imported plastic waste increased until 2012, and the recycling rate gradually increased along with the increase. However, some imported plastic waste could not be recycled because they were not sorted or they contained foreign matter. This has caused other problems, such as illegal dumping in China. With the rapid increase in illegal dumping of plastic waste and imported plastic waste in China,

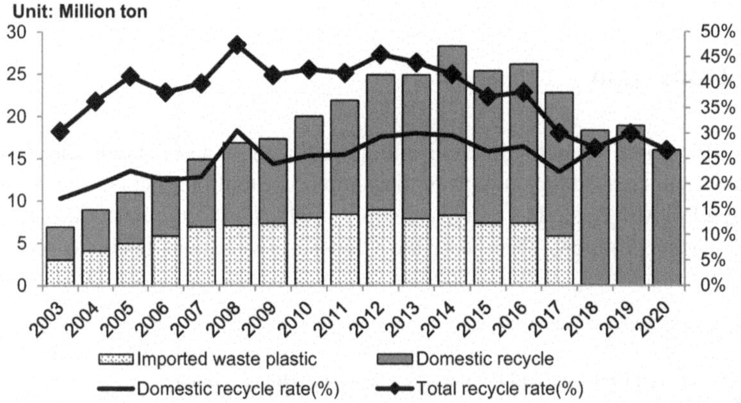

Fig. 10.1 Trends in volume and collection rate of plastic waste in China. Source: created by the authors based on the data from the Chinese Plastics Industry Yearbook (China Plastics Industry Association n.d.) and data from the Recycled Plastics Subcommittee of the Chinese Material Recycle Association

which is becoming a social problem, the Chinese government has been forced to take action. Against this backdrop, China's Ministry of Environmental Protection revised the "Imported Waste Management Inventory" in August 2017 to restrict the import of the plastic waste. [1] The amount of imported plastic waste decreased to 520 thousand tons in 2018, 1000 tons in 2019, and 0 in 2020. The sluggish recycling rate of the plastic waste in China is also a reflection of the restrictions on the importation of the plastic waste.

In China, uncollected plastic waste are disposed of as domestic wastes. According to the Chinese Statistical Yearbook, 80.4% of China's domestic wastes disposed were landfilled and 16.0% were incinerated in 2007. As the number of waste incineration facilities in China increases, the incineration rate of the wastes has also increased, reaching 58.9% in 2020. Even if a certain portion of the uncollected plastic waste is incinerated, a significant portion is landfilled. Because the plastics are toxic and take more than one hundred years to decompose naturally, landfilling the plastics may cause soil contamination. From another perspective, since the plastic waste have a large landfill volume, they exacerbate the problem of landfill site shortages.

3 Model and Data

At the International Symposium on Environmental Destruction held in Tokyo in 1970, Dr. Leontief argued for the need of an Input–Output table (IOT) for environmental analysis. Since various types of waste are generated in each industrial sector, it is necessary to create a waste input–output table (WIOT) that lists how the wastes are treated, disposed of, or recycled. The concept of the WIOT was given a concrete form by Dr. Nakamura who compiled Japanese WIOT in 1995 and 2000.

In this study, the WIOT for China is first estimated based on its IOT for the year 2007 (China National Bureau of Statistics 2009a, 2009b). Then, the environmental impacts of China's plastic waste control policy are estimated using this table. As an indicator of the environmental impacts, the impacts on nationwide reduction in landfill area and volume of waste are used.

3.1 What Is the WIOT?

The WIOT is a type of input-output table in which "venous" movements, such as a waste material flow and a waste treatment, are introduced to the usual IOT that

[1] Until then, both domestic and industrial plastic waste had been included in the "List of solid wastes whose importation is restricted." However, with this revision, "domestic plastic waste" are now included in the "List of prohibited imports," and "industrial plastic waste" are included in the "List of restricted imports."

describes "arterial" movements, such as a purchase, production, and sale of goods and services. The WIOT is a hybrid type IOT in which the physical unit table is added to the transaction table of the usual IOT.

The characteristics of WIOT can be summarized in the following three points according to Nakamura and Kondo (2009). First, in addition to the supply–demand equilibria of goods and services, the material balance of the wastes is also in equilibrium in the WIOT. In other words, the WIOT describes whether the wastes are finally treated or re-inputted into the industrial sector. Second, since the WIOT also covers the material cycle between industries, the environmental impact of the entire economic activity can be evaluated. Therefore, it is possible to estimate how the effect of a change in the recycling or treatment status of a certain waste material will spill over to the overall waste volume in the economy. Third, since the WIOT includes waste treatment sectors such as incineration, shredding, and landfill, it is also possible to analyze technological changes by treatment sector.

In this study, the Chinese version of the WIOT is developed within the framework of Nakamura and Kondo (2009). In China, waste disposal is broadly classified into landfill and incineration. However, the wastes are not always disposed of properly due to the shortage of the treatment facilities there. Thus, the Chinese version of the WIOT in this study includes a pseudo-industry, or "others," as part of the waste treatment sector, in addition to "landfill" and "incineration." Input to the "others" sector refers to the wastes that are not delivered to appropriate treatment facilities (landfills and incineration plants). That is, it corresponds to the illegal dumping.

Figure 10.2 shows a conceptual diagram of the ordinary IOT and the WIOT. The WIOT represents waste streams in addition to goods and services streams. The ordinary IOT and the WIOT share the same column sector, but there are differences in the row sectors. The ordinary IOT has a processing sector in both as the column sector and the row sector, while WIOT has a row sector for net waste discharge instead of a processing sector. The net waste discharge is calculated as the difference between the discharge and the input. The sum of the rows is the amount of each type of waste to be treated.

Subscript 1, 2, and f represent the industrial sectors, waste treatment sectors, and final demand sectors, respectively. Let n_1, n_2, and m be the number of industrial sectors, the number of treatment sectors, and the type of wastes, respectively. In the Input-Output Table, $\mathbf{X}_{1,\ 1}$ is an $n_1 \times n_1$ matrix representing the input from the industrial sector to the industrial sector, and $\mathbf{X}_{1,\ 2}$ is an $n_1 \times n_2$ matrix representing the input from the industrial sector to the waste treatment sector. Likewise, $\mathbf{X}_{2,\ 1}$ is an $n_2 \times n_1$ matrix representing the input from the waste treatment sector to the industrial sector, and $\mathbf{X}_{2,\ 2}$ is an $n_2 \times n_2$ matrix representing the input from the waste treatment sector to the waste treatment sector. $\mathbf{x}_{1,\ \mathbf{f}}$ represents the final demand of the industrial sector, while $\mathbf{x}_{2,\ \mathbf{f}}$ represents the final demand of the waste treatment sector.

The WIOT includes the net waste generation in the last row of the table. This row is expressed in physical quantity, not in monetary value. The total amount of the wastes is represented by a vector $\mathbf{w}$ corresponding to the types of waste. $\mathbf{W}_1$ is an $m \times n_1$ matrix of net waste discharge from the industrial sector, and $\mathbf{W}_2$ an $m \times n_2$ matrix of the net waste discharge from the treatment sector. The discharges from the

Ordinary Input–Output Table

	Industry sectors	Processing sectors	Final demands	Row total
Industry sectors (monetary unit)	$\mathbf{X}_{1,1}$	$\mathbf{X}_{1,2}$	$\mathbf{x}_{1,f}$	$\mathbf{x}_1$
Processing sectors (monetary unit)	$\mathbf{X}_{2,1}$	$\mathbf{X}_{2,2}$	$\mathbf{x}_{2,f}$	$\mathbf{x}_2$
Value added (monetary unit)	$\mathbf{V}_1$	$\mathbf{V}_2$		$\mathbf{v}$

Waste Input–Output Table

	Industry sectors	Processing sectors	Final demands	Row total
Industry sectors (monetary unit)	$\mathbf{X}_{1,1}$	$\mathbf{X}_{1,2}$	$\mathbf{x}_{1,f}$	$\mathbf{x}_1$
Net waste emissions (physical unit)	$\mathbf{W}_1$	$\mathbf{W}_2$	$\mathbf{w}_f$	$\mathbf{w}$
Value added (monetary unit)	$\mathbf{V}_1$	$\mathbf{V}_2$		$\mathbf{v}$
Environmental load (physical unit)	$\mathbf{e}_1$	$\mathbf{e}_2$	e_f	e

Fig. 10.2 Model of the waste input–output table (WIOT). Source: created by the authors based on Nakamura and Kondo (2009)

waste treatment sector are, for example, ashes generated during the incineration process.

However, since there is no one-to-one correspondence between waste types and treatment methods, the WIOT becomes a rectangular matrix. Thus, Nakamura and Kondo (2009) separately defined an allocation matrix to each waste treatment method to create a square table where the wastes were linked to the treatment sectors, and then transformed it into the IOT for which an inverse matrix could be defined. The analytical model by Nakamura and Kondo (2009) is explained below.

The amount of the wastes to be treated for each type of waste, i.e., vector $\mathbf{w}$, can be expressed by Eq. (10.1), where $\boldsymbol{\iota}_n$ is an $n \times 1$ vector with all elements equal to 1 (aggregate vector).

$$\mathbf{w} = \mathbf{W}_1 \boldsymbol{\iota}_{n1} + \mathbf{W}_2 \boldsymbol{\iota}_{n2} + \boldsymbol{w}_f \tag{10.1}$$

Next, let the input coefficient matrix representing the amount of goods and services input per unit of production be $\mathbf{A}_{1,\,1}$ and $\mathbf{A}_{1,\,2}$, and the waste discharge coefficient matrix representing the net waste discharge per unit of production be $\mathbf{G}_1$ and $\mathbf{G}_2$, to obtain a demand–supply equilibrium equation for the industrial sector (Eq. 10.2) and a demand–supply equilibrium equation for the wastes (Eq. 10.3).

$$\mathbf{x}_1 = \mathbf{A}_{1,1}\mathbf{x}_1 + \mathbf{A}_{1,2}\mathbf{x}_2 + \mathbf{x}_{1,f} \tag{10.2}$$

$$\mathbf{w} = \mathbf{G}_1\mathbf{x}_1 + \mathbf{G}_2\mathbf{x}_2 + \mathbf{w}_f \tag{10.3}$$

If the unknowns in the aforementioned equation system are the production values of the industrial and treatment sectors and the amount of wastes, there are $n_1 + n_2 + m$ unknowns, while there are only $n_1 + m$ equations, which means that unique solutions cannot be obtained. This is because the WIOT lacks a description of the waste flows, i.e., which waste is processed in which treatment sector.

The allocation matrix is defined to link each waste to the treatment sector. Let z_{ik} be the quantity of the k-th waste to be processed in i-th treatment sector, and $\mathbf{Z}$ be an $n_2 \times m$ with z_{ik} as its elements. Then, a matrix representing the distribution of treatments $\mathbf{S}$ is defined as in Eq. (10.4). The elements of the allocation matrix $\mathbf{S}$ is $S_{ik} = z_{ik}/w_k$, indicating the proportion of the waste treated in the i-th treatment sector in the total amount of each waste to be treated, w_k.

$$\mathbf{S} = \mathbf{Z}\widehat{\mathbf{w}}^{-1} \tag{10.4}$$

The production of the treatment sector (total throughput) $\mathbf{x}_2$ is defined by the following Eq. (10.5).

$$\mathbf{x}_2 = \mathbf{S}\mathbf{w} \tag{10.5}$$

Using Eq. (10.5) (correspondence between each waste and the treatment sector), the models for Eq. (10.2) (supply–demand matching equation for the goods and the services) and Eq. (10.3) (balance equation for the wastes) can be reconstructed so that both the number of unknowns and the number of equations become $n_1 + n_2 + m$; thus, the equations can be solved.

Considering the allocation matrix, the basic equation of the analytical model using the WIOT is given by Eq. (10.6).

$$\begin{bmatrix} \mathbf{x_1} \\ \mathbf{x_2} \end{bmatrix} = \begin{bmatrix} \mathbf{A_{1,1}} & \mathbf{A_{1,2}} \\ \mathbf{SG_1} & \mathbf{SG_2} \end{bmatrix} \begin{bmatrix} \mathbf{x_1} \\ \mathbf{x_2} \end{bmatrix} + \begin{bmatrix} \mathbf{x_f} \\ \mathbf{Sw_f} \end{bmatrix} \tag{10.6}$$

This equation is similar to the basic equation of an ordinary input–output analysis, and the equilibrium quantity of waste is obtained using Eq. (10.7).

$$\begin{bmatrix} \mathbf{x_1} \\ \mathbf{x_2} \end{bmatrix} = \left[\mathrm{I} - \begin{bmatrix} \mathbf{A_{1,1}} & \mathbf{A_{1,2}} \\ \mathbf{SG_1} & \mathbf{SG_2} \end{bmatrix} \right]^{-1} \begin{bmatrix} \mathbf{x_f} \\ \mathbf{Sw_f} \end{bmatrix} \tag{10.7}$$

Finally, when coefficients (vectors) representing the environmental impact per unit of production in the industrial and treatment sectors are $\mathbf{r_1}$ and $\mathbf{r_2}$, the environmental impact of the industrial and treatment sectors is defined as in Eq. (10.8).

$$[\mathbf{e_1} \ \mathbf{e_2}] = [\mathbf{r_1} \ \ \mathbf{r_2}] \begin{bmatrix} \mathbf{x_1} & 0 \\ 0 & \mathbf{x_2} \end{bmatrix} \tag{10.8}$$

3.2 *Structure of the Chinese WIOT*

The impact of promoting recycling activities is not limited to their own industry, but extends to other industrial sectors as well. For example, if the recycling rate of the plastics increases, the amount of chemicals used in the recycling process will increase, and the amount of plastic waste and chemicals transported by the transportation sector will also increase, which in turn increasing the environmental impact of each industry. Using WIOT, such effects on the environmental impact of other industries can be analyzed.

WIOT is composed of a basic transaction table (value), a net waste discharge table (quantity), an environmental impact table (quantity), and an allocation matrix table (quantity). The following sector outlines the estimation procedure for each table.

Basic Transaction Table

In the industrial classification of the WIOT, the waste management sector is required. However, in the 2007 Table [2] that had 135 sectors, the waste treatment sector was a part of the Water supply, environment, and public facilities management industry sector. Therefore, the waste treatment sector must be separated from it. The waste treatment sector (column sector) was divided into three sectors, namely landfill, incineration, and others, based on the current situation in China. For the industrial sectors other than the waste treatment sector, the 135-sector table was merged into a 45-sector table for analytical convenience. The basic transaction table (value table) that follows these steps is a 45 × 48 matrix.

Net Waste Discharge Table

The amount of waste discharged and recycled in each industrial sector by type was estimated by the authors. The net amount of waste discharged minus the amount recycled is the net amount of waste discharged.

Although this study focuses on the plastic waste for preparing the WIOT, the classification of the China National Development and Reform Commission (2014) was used as a reference, and a total of 21 types of wastes—17 types of industrial wastes and four types of domestic wastes—were covered in this study. The 17 types of industrial wastes are rice straw, agricultural films, fly ash, ore residues, coal gangue, desulfurized gypsum, waste ore, cinder, scrap metals, waste paper, plastic waste, hazardous wastes, other industrial wastes, incineration residues, incineration ash, construction wastes, and crushed dust. The four types of domestic wastes are scrap metals, waste paper, plastic waste, and other domestic garbage. Therefore, this part is represented by a 21 × 48 matrix in terms of quantity (in tons).

Environmental Impacts

Two environmental load factors in this study are landfill area and landfill volume. The amount of each environmental load factor in each industrial sector and waste disposal sector was obtained through the estimation by the author. This part is represented by a 2 (environmental load factors) × 48 matrix.

Allocation Matrix

The allocation matrix in this study is a 3 (treatment methods) × 21 (wastes) matrix. The treatment methods for each waste are identified, and their shares are estimated.

[2]The most detailed table among the Chinese IOT in 2007 is the one that had 135 sectors.

3.3 *Estimation of Chinese WIOT*

Estimation of Data Related to the Plastic Waste

Amount of the Plastic Waste Discharged and Collected (Agricultural, Industrial, and Domestic)

Nationwide consumption and discharge rates of plastic products in 2007 are obtained from the China Plastics Industry Yearbook 2008. Calculating the amount of the plastic waste discharged as Consumption × Discharge rate of the plastics, the total amount of the plastic waste discharged nationwide is estimated to be 16,526 thousand tons. Of those, the agricultural film [3] accounted for 1912 thousand tons, and the general plastics (films, pipes, synthetic leather, daily necessities, etc.) account for 15,614 thousand tons. In this study, the general plastic wastes were estimated by dividing it into industrial plastic waste and domestic plastic waste. However, since the nationwide allocation ratios were not available, the allocation ratio for Guangdong Province was applied to the entire country. The allocation method is described below. The total amount of the general plastic waste in Guangdong Province in 2008 was estimated to be 1615 thousand tons by multiplying the amount of plastic consumption (3135 thousand tons) by the disposal rate (51.5%) (Yang et al., 2012). Nationwide plastic waste collection rate in 2008 was 30.5% (China Plastics Industry Yearbook 2009), and the amount of plastic waste collected from households in Guangdong Province was 399.6 thousand tons (Yang et al., 2012). Therefore, using the relationship, Amount of plastic waste collected = Total amount of plastic waste × Collection rate, the total amount of the domestic plastic waste in Guangdong Province can be estimated to be 1310 (399.6/0.305) thousand tons. Next, the gross amount of the industrial plastic waste in Guangdong Province can be estimated to be 305 (1615 − 1310) thousand tons as the difference between the amount of the general plastic waste and the domestic plastic waste. As a result, the ratio of the industrial plastic waste to the domestic plastic waste is estimated to be 18.85:81.15. Assuming that the allocation ratios for the entire country and in Guangdong Province are the same, the nationwide amount of the industrial and the domestic plastic waste discharged are estimated to be 2943 thousand tons (15,614 × 18.85%) and 12,671 thousand tons (15,614 × 81.15%), respectively.

The amounts of industrial and domestic plastic waste collected (input for recycling) nationwide are estimated next. The averages of the collection rate of the plastic products (21.4%) and the agricultural film (80.3%) in 2007 can be obtained from the "China Plastics Industry Yearbook." From these data, the collection rates of the agricultural film, the industrial plastic waste, and the domestic plastic waste are estimated to be 80.30%, 17.96%, and 17.96% of the total amount of waste

[3] Agricultural films refer to plastic sheets for plastic greenhouses and tunnel cultivation, as well as plastic sheets for heat retention, moisture retention, insect control, etc.

discharged, respectively. However, for the industrial plastic waste, since China imported a large amount of plastic waste for recycling, the amount of the plastic waste imported in 2007 (6912 thousand tons) obtained from the China Customs Clearance Statistical Yearbook (General Administration of Customs of P. R. China (2008) was added to the input amount for recycling.

Amount of the Plastic Waste Discharged and Collected (Automobiles and Home Appliances)

In addition to the agricultural films, the industrial plastics, and the domestic plastics, the plastic waste from automobiles and home appliances discarded must be taken into account. The discarded automobiles and home appliances are types of industrial wastes, but are not included in the data from the Chinese Plastics Industry Yearbook 2008. Therefore, they need to be estimated separately and added to the amount of industrial plastic waste discharged and collected. The processing (dismantling and sorting) of discarded automobiles and home appliances is to be handled by recyclable resources recovery and processing operators. The recyclable resources recovery and processing operators input the recycled resources obtained from dismantling operations to other industrial sectors as intermediate input goods. However, since it is impossible to know the amount of illegally dumped automobiles and home appliances, this study assumed that there was no illegal dumping of the discarded automobiles and home appliances.

Hiraiwa (2011, p.83) estimated the number of the discarded automobiles in China as Number of vehicles owned at the end of the previous year + Number of vehicles sold during the current year − Number of vehicles owned at the end of the current year. In this study, the data on the number of vehicles owned in 2006 and the number of vehicles sold and owned in 2007 obtained from the China Industrial Statistical Yearbook 2008 were used to assume the number of the discarded automobiles in 2007 of about 2580 thousand. The average weight of a scrapped vehicle was assumed to be about 3.3 tons according to Mid- to Long-term Plan for Recyclable Resources Recovery System Development (2015–2020). According to China Ministry of Commerce and China National Resources Recycling Association (2012), the amount of resources recoverable through the dismantling of the discarded automobiles is 1.8 tons, with the remainder being crushed dust. The breakdown of the recovered resources is as follows: steel scraps, 72.0%; nonferrous metal scraps 6.0%; plastic waste, 6.3%; waste rubber and glass, 6.1%; waste oil, 1.0%; and other materials, 8.7%. Therefore, by multiplying each of the obtained values by the number of vehicles in 2007, the amount of waste discharged from the discarded automobiles by type can be estimated ("Discarded vehicles" column in Table 10.1). Of those, the amount of the plastic waste was about 293 thousand tons.

Meanwhile, as for the discarded home appliances, those for which the waste can be estimated from statistical data include five product groups: washing machines, refrigerators, televisions, air conditioners, and personal computers. According to Liu et al. (2005, pp.114–115), the average useful lives of these home appliances are

Table 10.1 Amount of waste from discarded automobiles and home appliances

		Discarded vehicles	Discarded home appliances					
			Washing machine	Refrigerator	TV set	Air conditioner	Personal computer	Home appliance total
Disposal quantity (1000 units)		2,580	11,620	10,190	43,120	11,460	13,680	
Average weight (ton/1000 units)		3,296	22	32	22	32	6	
Gross weight (1000 tons)		8,503	256	326	949	367	82	1;979
Discharge (1000 tons)	Waste plastic	293	77	131	224	55	16	503
	Scrap metal	3,618	135	165	138	278	29	745
	Crushed dust	3,865	21	9	500	10	24	565
	Others	727	23	20	86	23	14	166

Note: "Others" refers to waste rubber, waste glass, etc.
Source: Prepared by the authors

10 years for washing machines, 10 years for refrigerators, 8 years for TVs, 11 years for air conditioners, and 5 years for PCs. Therefore, it can be assumed that the quantity of discarded home appliances in 2007 is equal to the number of appliances sold in the year that goes back by the average service life of each product. For example, in the case of the washing machine with the average service life of 10 years, the number of units discarded in 2007 is assumed to be the number of units sold in 1998, i.e. 10 years earlier. The sales volume of the home appliances can be obtained from the Chinese Industry Yearbook. Similarly to the discarded vehicles, the discarded home appliances become either the recovered resources or the shredded dust. Yang et al. (2012) and Liu et al. (2005, p.116) estimated their average weights from the discarded home appliances and material ratio of each type. This study used their estimates. By multiplying the number of items discarded by the average weight and material ratio, the amount of waste discharged by waste type for each discarded home appliance was estimated. ("Discarded Home Appliances" in Table 10.1). For the wastes recovered as resources, the amount discharged and the amount input were the same. For the shredded dust, only the discharges were considered, and no inputs to other industries were assumed. As a result, the total amount of plastic waste discharged from the home appliances was about 503 thousand tons.

To summarize, the amounts of the plastic waste discharged were 912 thousand tons from the agricultural films, 293 thousand tons from the industrial plastic waste from the discarded automobiles, 503 thousand tons from the industrial plastic waste from the discarded home appliances, 2943 thousand tons from the industrial plastic waste from other industries, and 12,670 thousand tons from domestic plastic waste. The amounts of the plastic waste inputs were estimated to be 732 (912 × 80.30%) thousand tons for the agricultural film, 8237 (2943 × 17.96% + 6912 + 293 + 503) thousand tons for the industrial plastic waste, and 2276 (12,670 × 17.96%) thousand tons for domestic plastic waste.

Allocation of the Plastic Waste by Column Sector

Since the agricultural film (910 thousand tons) is an agricultural waste, its discharge is recorded in the agriculture, forestry, and fisheries sector. Its input is recorded in the plastic products manufacturing industry sector because the discarded agricultural film is used for plastic recycling.

There are no statistical data available on the amount of the plastic waste discharged by sector other than the agricultural film. Therefore, it was assumed that the plastic wastes were discharged proportionally to the input of the plastics. According to the China National Bureau of Statistics (1996), there were four [4]industrial sectors that did not use the plastics: oil and natural gas extraction,

[4]Include passenger cars and freight vehicles (both for public and private use). However, public transportation vehicles such as city buses and special work vehicles such as mining vehicles are not included.

Table 10.2 Allocation matrix for Chinese WIOT

Treatment sector	Landfill	Incineration	Others (illegal dumping)
Wastes			
Agricultural film	0.5473	0.1029	0.3498
Industrial waste plastic	0.5473	0.1029	0.3498
Household waste plastic	0.5025	0.0945	0.4030

Source: Prepared by the author

extraction non-metallic and other minerals, accommodation and catering, and finance. Thus, 3739 (293 + 503 + 2943) thousand tons of industrial plastic waste were allocated according to the amount of plastic input into 41 industrial sectors excluding these four sectors. On the other hand, the input of the industrial plastic waste (8237 thousand tons) is the amount that was collected and recycled. Therefore, all of their input was accounted for in the plastic manufacturing industry sector. The input of the domestic plastic waste (2276 thousand tons) was allocated to the household sector (final demand).

Estimation of the Allocation Matrix for Plastic Waste

As mentioned above, the waste allocation matrix was used to link the waste to the waste disposal to convert the WIOT into a square matrix. In the model in this paper, there are three waste treatment sectors: landfill, incineration, and other (effectively, illegal dumping). The allocation matrix for the plastic waste was created as shown in Table 10.2.

Estimation for the Waste Disposal Sector

Since there were no official statistics on the waste processing costs for landfill and incineration, this study used the study on waste disposal costs for landfill sites and incineration plants by Shi et al. (2012). Since these are included in the Water Conservation, Environment and Public Facility Management, we divide this industry.

The following procedure was used to estimate the amount of the waste processed by each waste treatment sector. First, using the allocation ratios of the waste treatment sectors, the amount of wastes processed by type obtained in the previous section was allocated to the amount of wastes landfilled and the amount of wastes incinerated. Then, by multiplying the processing cost per ton waste by the total amount landfilled and the total amount incinerated, the inputs for the landfill incineration sectors were estimated. Finally, the Water Conservation, Environment and Public Facility Management industry except Landfill and Incineration is divided by subtracting the total input of the waste treatment industries (Landfill and

Incineration) from the Water Conservation, Environment, and Public Facility Management industry in the IOT.

Estimation of Environmental Load Factors

The landfill sector uses land for landfill disposal of the waste. Additionally, the Other (illegal dumping) sector uses land in forests and cities where the wastes are illegally dumped. Therefore, the land used by the waste dumped in the Other (illegal dumping) sector was considered in the same way as the land used in the Landfill sector in this study. The square footage of landfill space and the volume of landfill waste used in both Landfill and Other (illegal dumping) sectors are estimated as follows.

The Square Footage of the *Landfill* Sector

Since the official statistics for landfill sites were not available for 2007, the values in 2007 were estimated based on the 2013 data. According to China Association of Urban Environmental Sanitation (2014), there were 1549 landfills in operation in 2013, with a total area of 36,262 thousand m^2. Based on the total area and the number of landfills, the average area of a single landfill site is ~23 thousand m^2. Multiplying this average area by the number of landfills in operation in 2007 (2135), the square footage of the landfills in operation in 2007 was estimated to be 49,980 thousand m^2.

The *Other* sector is a hypothetical sector (illegal dumping sector) for which no statistical data were available; however, it was assumed that the land used by this sector was the same as that used by the *Landfill* sector. In this study, the allocation ratios of the amount discharged, the amount of input and the disposal methods for each waste type were estimated not only for the plastics but also for 21 waste types, which can be used to estimate the amount of landfill for both the *Landfill* and *Other* sectors by waste type. The total landfill volume in the *Landfill* sector obtained was then divided by the area of landfill sites, and the area of landfill sites occupied by the wastes per ton was estimated to be 51 m^2. Multiplying this figure by the total amount of the wastes dumped in the Other sector (illegal dumping, 595,350 tons), the land area of its dumping sites was estimated to be 30,363 thousand m^2.

Landfill Volume of the Landfill Sector

The landfill volume is obtained by dividing the total volume of the wastes by the bulk density and compaction factor. Based on the reports of Hokkaido University (1998) and Xu (2010), the bulk density of the plastic waste at the time of collection was assumed to be 0.21 t/m^3 and the compaction coefficient was considered to be 4.05 (Table 10.3). In the Landfill sector, the wastes are crushed and compacted.

Table 10.3 Landfill volume by type of plastic waste (Unit: 1000 m3)

	Landfill	Illegal dumping	Total
Agricultural film	116	300	416
Industrial waste plastic	1,535	3,974	5,510
Household waste plastic	6,142	19,949	26,091
Total	7,798	24,224	32,016

Source: Prepared by the author

Table 10.4 Assumption on the plastic waste collection rates

Scenario	Agricultural film	Industrial waste plastic	Household waste plastic
Base	80.3%	18.0%	18.0%
Scenario 1	80.3%	27.1%	27.1%
Scenario 2	85.0%	27.1%	60.0%
Scenario 3	100.0%	30.0%	100.0%

Source: Prepared by the authors

Therefore, their volume is smaller than that when they were collected. However, the illegally dumped wastes remain in the same state as when they were collected. In the case of plastic waste, the bulk density at the time of collection was assumed to be the same as in the Landfill sector (0.21 t/m^3), but the compaction factor was assumed to be 1.0 (Table 10.3).

4 Simulation Analysis and Results

4.1 *Setting up Simulation Scenarios*

The following three simulation scenarios were assumed to evaluate the effects of the recycling promotion policy on the reduction of final disposal volume. Since there are three types of plastic waste in this study, namely agricultural film, industrial plastic waste, and domestic plastic waste, the assumed collection rates of the plastic waste in each simulation scenario are shown in Table 10.4.

Scenario 1: Assuming an Increase in the Collection Rate of the Industrial Plastic Waste and the Domestic Plastic Waste

The plastic waste recycling industry has received various preferential treatments, including value-added tax exemptions, and the collection rate of the plastic waste has increased considerably since 2007. Based on this situation, Scenario 1 assumed that the collection rate of the industrial and the domestic plastic waste increased by 1.5 times, from 18.0% in 2007 to 27.1%. However, the collection rate of the agricultural film was assumed to remain unchanged at 80.3%.

Scenario 2: Assuming Fulfillment of the Latest Plastic Waste Collection Policy Targets

Using the target values in the Fourteenth Five-Year Plan, this scenario assumed that the collection rate of the agricultural film and the domestic plastic waste would increase to 85.0% and 60.0%, respectively, by 2025.

Scenario 3: Assuming the Best Case in the Future

The reason why the collection rate of the plastic waste has not increased much is due to the difficulty of collecting industrial plastic waste. On the other hand, it will be possible to increase the collection rate of the domestic plastic waste to 100% in the future. In addition, the Regulations Governing Agricultural Films, which took effect in 2020, calls for a 100% collection rate for the agricultural films. Therefore, it should also be possible to achieve a 100% collection rate for the agricultural films. Thus, this scenario assumed that the collection rates of the agricultural film and the domestic plastic waste would become 100%, while the collection rate of the industrial plastic waste would remain at 30%.

4.2 *Estimation of Net Waste Discharge by Simulation*

The simulation of the improvement in the collection rate of the plastic waste is to make changes to waste discharge factor matrices of the industrial and the waste treatment sectors $\mathbf{G_1}$ and $\mathbf{G_2}$. For this purpose, the net waste discharge is estimated. As the amount of the plastic waste recycled increases, the net waste discharge decreases. The incineration sector of the waste treatment sector discharges incineration residue and ash. As the amount of incineration decreases, the amounts of incineration residue and ash discharge also decrease, and their input amounts to the ceramic, stone, and clay industry decrease accordingly. As a result, net discharge from the ceramic and stone industry (i.e., discharge minus input) increases slightly. However, the amount of discharge suppressed by the incineration sector is much larger than the increase in net discharge from non-ceramics. The change in net discharge is shown in Table 10.5. The discharge factors were modified based on the obtained net discharge.

4.3 *Analysis Results and Discussion*

The simulation results are summarized in Table 10.6. The reductions in landfill square footage and volume were linked to the increase in the collection rate of the

Table 10.5 Changes in waste discharge in the simulation (Unit: 1000 tons)

		Industrial sectors		Waste treatment sectors		
		Plastic	Cement	Incineration	Others	Total
Scenario 1	Industrial waste plastic	−268	0	0	−94	−362
	Incineration residue	0	11	−28	0	−17
	Incinerated ash	0	0	−6	0	−6
	Household waste plastic	−1,153	0	0	−465	−1,618
Scenario 2	Industrial waste plastic	−1,012	0	0	−354	−11,366
	Incineration residue	0	43	−106	0	−63
	Incinerated ash	0	2	−23	0	−21
	Household waste plastic	−4,358	0	0	−1,756	−6,114
Scenario 3	Industrial waste plastic	−2,060	0	0	−721	−2,781
	Incineration residue	0	88	−216	0	−128
	Incinerated ash	0	3	−46	0	−43
	Household waste plastic	−8,869	0	0	−3,574	−12,443

Source: Prepared by the authors

Table 10.6 Estimation of the reduction in waste landfill volume by scenarios

	Landfill area (1000 m^2)			Landfill volume (1000 m^3)		
	Landfill	Others	Total	Landfill	Others	Total
Scenario 1	−402	−294	−696	−499	−479	−978
Scenario 2	−1,518	−1,112	−2,630	−1,884	−1,811	−3,696
Scenario 3	−3,089	−2,263	−5,353	−3,835	−3,686	−7,522

Source: Prepared by the authors

plastic waste, indicating that the promotion of recycling is an effective solution to the problems concerning the wastes.

Scenario 1 assumed a 1.5-fold increase in the collection rate of the plastic waste, and this resulted in the reductions of 696 thousand m^2 of landfill area and 978 thousand m^3 of landfill volume. However, this is only about 1.4% of the actual landfill operations in 2007. Considering the fact that the annual growth rate of waste volume in China from 2008 to 2019 was about 5%, it cannot be denied that the effect of the increase in the plastic waste collection rate has been extremely limited.

Scenario 2 assumed that the government targets for 2025 (increase in the collection rate of the agricultural film to 85.0% and increase in the collection rate of the domestic plastic waste to 60.0%) would be achieved. A significant increase in the collection rate of the domestic plastic waste is expected to save 2630 thousand m^2 of landfill area and 3696 thousand m^3 of landfill volume. [5] Although this scenario is expected to result in relatively large reductions, it represents only 5.3% of the actual

[5] Since the area and the volume of the Tokyo Dome in Japan is 46 thousand m^2 and 1240 thousand m^3, respectively, the savings in the landfill space and the final waste disposal volume are equivalent to 56 and 3 Tokyo Domes in terms of area and volume, respectively.

landfill volume in 2007, an amount that corresponds to a one-year increase in the volume of waste.

Scenario 3 assumed that the domestic plastic waste and the agricultural film would be fully collected. In this case, savings were estimated at 5352 thousand m^2 of landfill area and 7521 m^3 of landfill volume.[6] This figure corresponds to 10.7% of the actual landfill volume in 2007.

5 Conclusions and Future Challenges

While plastic consumption continues to increase in China, its collection rate of domestic plastic waste remains at a low level, and the amount of plastic waste is increasing. Since secondary waste is discharged in each industrial sector in the process of recycling plastic waste, this study estimated the Chinese version of the Waste Input–Output Table (WIOT) and evaluated the effect of promoting the recycling of the plastic waste on the reduction of final waste disposal amount. The conclusions obtained from the analyses in this study can be summarized in the following two points. First, the promotion of plastic waste recycling is effective in alleviating the shortage of final waste disposal sites. Second, the current subsidy and preferential policies for the recycling of plastic waste have limited impacts. To improve the collection rate of the plastic waste, it is essential to further strengthen these policies, raise citizens' awareness on sorting by resources, and increase the number of collection sites. In addition, with the assumption of this study, the amount of illegal dumping increases proportionally as the amount of plastic waste increases. From the standpoint of environmental preservation, it is necessary to take measures to control the illegal dumping.

The biggest remaining challenge for this study is data preparation. The Chinese version of the WIOT used in this study is provisional and does not reflect recent technology. Due to limitations of the available data, regional, and annual consistency of the data was not necessarily achieved. The improvement of data on waste in China is expected, and the Chinese version of WIOT needs to be updated.

Acknowledgement This work was supported by JSPS KAKENHI Grant Numbers JP19K12459, JP20K12291, and JP21H04941.

References

China Association of Urban Environmental Sanitation (2014) China urban environmental and sanitation industry development report 2013. (in Chinese)

[6] As in Scenario 2, the figures are 114 and 6 Tokyo Dome equivalent in terms of area and volume, respectively.

China Ministry of Commerce and China National Resources Recycling Association (2012) Analysis report on the scrap vehicle recycling industry
China National Bureau of Statistics (1996) China physical input-output table 1992. China Statistics Press. (in Chinese)
China National Bureau of Statistics (2008) China industrial statisitical year book 2008. (in Chinese)
China National Bureau of Statistics (2009a) China input-output table 2007: compilation methodology. China Statistics Press. (in Chinese)
China National Bureau of Statistics (2009b) China input-output table 2007. China Statistics Press. (in Chinese)
China National Development and Reform Commission (2014) Annual report on utilization of resources in China 2014. (in Chinese)
China Plastics Industry Association (n.d.) China plastics industry yearbook (2008–2013). China Petrochemical Press. (in Chinese)
General Administration of Customs of P. R. China (2008) China customs yearbook 2007. China Customs Press. (in Chinese)
Hiraiwa Y (2011) Study on the number of end-of-life vehicles in China. In: Masaaki K (ed), Recent developments in environmental economics, (Institute of Economics Research, Discussion Paper Series B No. 40), 73–90. (in Japanese)
Hokkaido University, Graduate School of Engineering (1998) *Research on development of evaluation calculation system to support comprehensive management of municipal solid waste*, Department of Waste Disposal Engineering, Graduate School of Engineering, Hokkaido University. (in Japanese)
Liu X, Yang J, Wang R (2005) Estimates of electronic product wastes in China. China's Popul Res Environ 15(5):113–117. (in Chinese)
Nakamura S, Kondo Y (2009) Waste input-output analysis: concepts and application to industrial ecology. Springer
Shi X, Zhang X, Chen Y, Ding Y, Lu Q (2012) Cost analysis and control of waste landfill: taking the disposal of domestic waste in Changzhou as an example. Environ Biotechnol 20(1):37–39. (in Chinese)
Xu H (2010) Disposal of domestic waste in villages and towns. China Construction and Industry Publishing. (in Chinese)
Yang Z, Xiao D, Yuan J (2012) A research on renewable resource industry from the viewpoint of industrial ecology. Science publisher. (in Chinese)

The manufacturer's authorised representative in the EU is Springer Nature Customer Service Centre GmbH, Europaplatz 3, 69115 Heidelberg, Germany. If you have any concerns regarding our products, please contact ProductSafety@springernature.com

Printed and bound by CPI Group (UK) Ltd, Croydon, CR0 4YY
01/10/2026
02225896-0019